HOW

TO

Mill

YOUR

GOLD & SILVER

BY

Hank Chapman, Jr.

© Copyright 2015

**A Complete Plain English Guide
For The
Amateur or Professional**

DEDICATION

Again, to my wife, Sue, for her patience, and to all those folks who have sought consultation on this subject for many years, and have asked for this book. Also, to my many absent family and friends who have gone to that Big Gold Mine in the Sky. May they Rest In Peace.

Hank Chapman, Jr.,
2795 Avenida Grande
Bullhead City, AZ 86442
928-758-2155
E-mail hchap@suddenlink.net

Published and Distributed By:

Sylvanite Publishing
www.sylvanitepublishing.com

Cover art of the old Moss Mine head frame by Sue Chapman. The head frame is currently on display in Bullhead City, AZ Community Park. It is thought to have been built in the 20's.

FOR INFORMATIONAL PURPOSES ONLY!

Since the author and publisher of this book have no control over the way the methods and procedures outlined in this book are used, this information is presented for **Informational Purposes Only!** Unless the author is standing beside you as you use this information, you are on your own! Please be careful!

The Publisher, Author and Retailer assume no liability whatsoever for any method in which this information may be utilized, or damages arising from the use of this information.

All chemicals, at a certain concentration, can be harmful to human life, or fatal. People die every year drinking a glass of water. Our bodies have to have water to live, yet water can easily kill you in the right circumstances. Read the information on personal protective equipment carefully, and use the appropriate protective equipment.

Never mix chemicals unless you are absolutely sure how they will react. Never breathe dust from dry chemicals. Never breathe dust from silica. Never breathe acid fumes. Store any containers of wet chemicals or chemical solutions in a ventilated cabinet. All of these fumes and dusts will damage your lungs, and can cause silicosis or chemical pneumonia. Open or closed chemical containers will leak corrosive fumes that will damage electronic equipment, and cause corrosion.

Mining and milling are industrial processes, and tragically, people are killed every year in this industry. Be aware of the hazards of the equipment you are using, and if you don't know what it does, stay away from the machine, or chemical in question. Even three man operations are regulated by the respective state that the operation is in. Learn the regulations, and follow them. Always dispose of all hazardous waste according to local, State and Federal regulations. Ascertain if permits are required for your operation, and if so, obtain the necessary permits.

If you don't know or understand what you are attempting to do, don't!

The safety rules are for your protection. Use them!

Table of Contents

Introduction:

The purpose of this book is to provide the information you will need to successfully mill most types of ores. There is a step by step process that must be followed for you to have the information you need. Don't skip the steps, or you'll pay the price for your error. Most people inexperienced in the mining industry do not understand why some things are done over and over, such as the continuous monitoring and assaying of the process (mill) stream as ore passes through the circuit. Or the constant assay of blast hole samples.

Ask yourself what values you have, what values you have recovered, what values you have lost, and what values are ahead of you. There's only one way to find out.

One of the most common errors are failure to maintain the analysis of the values in the mill circuit. An old adage in the mining industry is "one assay does not make a mine". Many hundreds of assays, perhaps, or thousands of assays will determine if a mine can be made, and then continuously operated at a profit.

Another serious problem is understanding classification, and milling or grinding to the point of mineral liberation. There are solid principles for these processes, and they have been around for a long, long time.

This book is written with the understanding that you have a pile of ore, or perhaps a dump stock, or even an old mine dump you have discovered that needs to be milled. Or maybe you have discovered a really nice bench placer. A "dump stock" is exactly what it sounds like. Or how about a few thousand tons of mill tails?

There are a lot of expensive steps you will have to take to ascertain the parameters for a successful (profitable) recovery of the values. Think like the old adage who, what, when, where and why. You will have to answer a lot of other questions as well. Try to "poor boy" your operation and you will have trouble from the beginning, and will probably fail.

In the last 25 years, your author has consulted for many mining operations in several different countries, and all over the western United States. During this time, many people have called or came to the author's house just to ask questions.

Sadly, most of them simply didn't have or couldn't get the resources they needed to finish the project. Some had bought useless equipment they didn't need, or had a "secret process" they paid a small fortune for that didn't work. Some of them were operating on the "one rock assay". Some of them even mortgaged their homes and other family members' homes for financing. That is a really, really bad idea.

The first chapter is on reference books. Check out the books listed there. You will be referred to them throughout the text so you will understand what is in those books that you need to know. Some are out of print and very hard to find, but at least put forth the effort to learn a little about an industry that is more complex than you know at this point. Some have been reprinted and are very reasonably priced.

Another piece of advice is that you seek the services of a competent mining engineer. These guys know about every phase of mining and most of them are pretty competent geologists as well. They will keep you out of trouble and can provide information on equipment you will need, not to mention the actual mechanics of mining.

Always keep the Internet in mind. Also remember "Caveat Emptor" let the buyer beware. There is a fairly large percentage of the information on the Internet that is just pure crap, structured purely to sell you something you don't need. Be especially careful when some of these yahoos claim miracle results and create a U-Tube video. It seems that the more technical information gets, the more unsafe these people become. Don't become a victim yourself. Make sure you understand what you are doing, and why you are doing it.

An old, abandoned house on Spruce Mountain, Nevada.

Chapter One

Reference Books

A Personal Introduction for Books

The information written on these pages did not spontaneously appear in my brain. I learned by reading the literature available on the subject. And yes, you can say I was fortunate to work in the industry for many years, and learned a lot by doing so. The pay in the mining industry is also really great. That helps when you have to buy books.

I've always been curious, and therefore have searched for more information on the various subjects covered in this book. Over the years, I've accumulated a decent reference library, and I have always had a better library than any lab or metallurgical facility I've worked at.

The point, I guess, is that this knowledge is pretty specialized, sometimes hard to find, and generally expensive. If you are serious about your enterprise, you need to know what you are doing, and please, please educate yourself! A lot of mining can be quite dangerous, and you can cover your tender butt just by a having basic knowledge of what you want to accomplish.

As an example, you are sent into a huge cyanide tank (empty) to descale it with dilute HCL, hydrochloric acid. I have a couple of acquaintances that did just that, and were telling me about it. Both are disabled. The cyanide that evolved from the scale when the HCL was applied destroyed their kidneys. They are lucky to be alive. They had no idea what they were getting into, and found out the hard way. A good general first year chemistry textbook would have saved them a lot of agony.

And yes to Doc Kellenbarger, an old friend and lab weenie, Avogadro's number is 6.0221415×10^{23}. Sorry, readers, you'll just have to look it up. Chemistry trivia, as it were. Yes, my chemistry gets rusty, that's why I have the books, so I can look it up.

So build a library, and have the information at your fingers. You'll be glad you did.

General Information

There are hundreds of books out there that pertain to the subject matter at hand. Some are available on the internet, and can be downloaded free, and many, many more can be purchased or ordered online. Obviously, there are a great many sites devoted to this subject, and most are good reading, some are crap, and some are just plain dangerous. Verify any information you get that you intend to use, especially any legal or technical information.

As an example, in the book "How To Smelt Your Gold And Silver" it is plainly stated that manganese dioxide is the optional component of a gold smelting flux. Somehow, a reader thought that "magnesium dioxide" was the correct chemical, and was trying to "oxidize" magnesium to a "dioxide" by melting it in a furnace. That is a really, really bad idea. Understand what you read, and verify your understanding. Don't get yourself or someone else injured.

Web Sites For Mining Books

Here is a list to get you started.

Action Mining Catalog: www.actionmining.com
Miner's Inc: www.minerox.com
Mining Books.com: www.miningbooks.com
Make Your Own Gold Bars: www.makeyourowngoldbars.com
American Society For The Advancement of Technology: (ASAT) Possibly out of business, no response online, phone, or e-mail. www.asat.volant.org
Legend, Inc: www.legend-reno.com

A Selected List

Here are some of the books you should have. The primary focus here is on the reference materials for you that will help your operational problems.

CRC Handbook of Chemistry and Physics

You really don't need the latest issue, buy a used one. The parts you are after haven't changed much over the years. A lot of this book is available online, and downloadable as a PDF file. This is a really intense book, about three inches thick. Understand that you are after two parts. The first section of interest is "The Elements". This section describes in great detail everything known about the elements, such as gold and silver. The second section you are after is "The Physical Constants of Inorganic Compounds". This section describes all the compounds of all the elements, such as Au_2CL_6, a claret red crystalline powder, very soluble in hot water. That's right...Water soluble gold. Sometimes you don't need chemistry. Identify your mineral, and read on.

Standard Methods of Chemical Analysis, N. H. Furman, Editor

There are five volumes in this set, however the first volume will cover everything the aspiring chemist or assayer would want dealing with chemical analysis of minerals. There is also a very good section on the Fire Assay. There are analytical techniques for all the elements here. This book is probably out of print, however there are some on the web as government surplus.

The Metallurgy of Gold, and The Metallurgy of Silver, by Sir T. K. Rose

Anything and everything about gold and silver can be found in these two books. The information is quite extensive, and if you think the old timers didn't know what they were doing, well, these books will change your mind. The books cover all aspects of gold and silver processing and equipment through the 1930's. You will notice a lot of the old equipment was taken out of these books, and renamed, somewhat modified, and is in use today.

LaRune's Rockpecker Notes by T. D. LaRune

This book is actually in four parts, Rockpecker, Goldpecker, Mineral Recognitions, and Chemical Testing. A great book for the prospector, with a lot of good pointers for the beginner. A good read, and still available at the time this was written.

Rare Metals Handbook, Second Edition, Edited by Hampel

If you're into some element beside gold, this is the book for you. This book covers everything but precious metals, including rare earths. There is also a flow chart for process for the element being described. Lots of good technical information and circuit descriptions.

The Extractive Metallurgy Of Gold, by J. C. Yannopoulos

This book deals with contemporary gold leaching with cyanide. Lots of good information on current practice in the mining industry. Stripping carbon, CIL, CIP, it's all here. This is a recent book, Mr. Yannopoulos was employed at Newmont Mining when it was written.

Fire Assaying, by Shepard and Dietrich

This book has been reprinted, and is available at most of the web sites previously listed. If you are at all interested in the fire assay, this book should be your first choice. A lot of people buy Bugbee's "A Textbook Of Fire Assaying", and feel like they are trying to decipher Greek. Read Shepard and Dietrich first, then try Bugbee.

Handbook Of Mineral Dressing by Taggart

This book is the Bible of grinding ores and other industrial minerals. It is a complex book, with methods for every type of processing. It is an older book, published by Wiley back in the day, and is now extremely hard to find. The second printing was in 1947. If you see one under a hundred dollars, buy it, they are also collector's item. Your author has had this book stolen, so there must be something worthwhile in it. Your author has replaced the stolen copy. Get yours. An excellent reference. If all else fails, see the SME Handbooks on the next page.

Recovery And Refining of Precious Metals by C. W. Ammen

This is an excellent reference book. There are a lot of chemical procedures, chemical tests, and information dealing with lab techniques. Apparently this book has been recently revised, so make sure you get the latest edition. It is easy to understand, and explains a lot of general chemical terms, as well. There are lots of illustrations, and a good glossary.

Mining Chemicals Handbook by Cyanamid

This book is gold, as far as mining chemistry goes. Flotation, leaching, flocculation, even mineralogy. The book is difficult to find. The old version is spiral bound, the newer one is a blue hardcover book. The book deals with all the mining chemicals produced by American Cyanimid. There is a lot of good chemistry as well.

Anatomy of a Mine from Prospect to Production by the US Department of Agriculture (Forest Service)

General Technical Report INT-GTR-35. If you are new to mining, or contemplating an operation, especially on Federal land, this is where you should start. There are sections covering Mining Law, Prospecting, Exploration, Development, Production, and Reclamation. All from a government land manager's perspective. It is downloadable from the web, and is discussed in further detail in this book. Go to www.fs.fed.us/rm/pubs_int/int_gtr035.html or Amazon.com. Amazon will charge for the report.

SME Mineral Processing Handbook, Volume 1 & 2, Edited by Weiss

These books pretty much replace the previously described "Handbook of Mineral Dressing" by Taggart, so don't despair if you can't find Taggart. These books cover as much as Taggart, and are more current. They are published by the Society of Mining Engineers.

Condensed Chemical Dictionary by Hawley

This book is handy when you need to look up a chemical, and find out more about it. Also has the chemical formulas for all the chemicals. If you are going to leach with any chemical, you'll need this one.

Chemical Technician's Ready Reference Handbook by Shugar & Ballinger

This book takes you through basic lab procedures, glassware, safety, lab math, and on and on. How to be a lab technician, and what to do, pretty much. Do you know how to fold a filter paper? It's in here. If you're doing any wet chemical work, you need this book.

Introduction To Chemical Principles by Stoker

This book is basically a Chemistry 101 text, and this one is not an absolute necessity. Your author has three of these "introduction" type chemistry books, two are the 101 type, the other is a 202 type book. The idea is to familiarize yourself with some chemical procedures so that you can safely work with chemicals. Understand what you are dealing with. You will also find Avogadro's number in these books.

Analytical Chemistry For Technicians by John Kenkel

This book is a must if you are interested in analytical chemistry. The book covers all the methods and equipment used in analytical work, as well as maintaining a notebook, titration, and most anything analytical. A lot of good information on the elements, as well.

Gold - History and Genesis of Deposits by Boyle

This book deals with all the different types of gold deposits and how they were formed. The book also covers placer deposits. The book covers all the theories of how gold deposits are formed. A very good historical representation of gold deposits from early times to current times. A good "rock" book.

The Geology of Ore Deposits by Guilbert & Park

Another good "rock" book. If you are on a quest to learn how the deposit you are interested in was formed, this is the book for you. Many diagrams, some photos. Information on quite a few mining districts around the world.

The Complete Book of Rocks and Minerals by Chris Pellant

This book covers all sorts of rocks and minerals, and has excellent high quality photos of them. There are descriptions of how rocks are formed, and the identification of the rocks and minerals is made easy by the photos. Also included for each mineral are tests for identifying each mineral. Very comprehensive "rock" book.

How To Smelt Your Gold And Silver by Hank Chapman, Jr.

If you are at all successful, you'll need this book to learn how to smelt your concentrate or placer fines to a Dore' bar. This book has all you need to know, flux recipes, equipment lists, plenty of photos and safety information. After the completion of the book you are reading, the smelting book will be revised. Contact the publisher or author for further information.

The Nalco Water Handbook Edited by Frank M. Kemmer

The definitive book on water, and the applications that use water. Plenty of photos, diagrams and drawings of treatments, processes and specialized treatments of water. Nalco is known worldwide for their water treatment chemicals, and this book is a description of how those chemicals are used, and a complete review of all the machinery used in the process. The third edition is available.

Advanced Dredging Techniques, Volumes 1 & 2 by Dave McCracken

If you are involved in placer mining, these books are a very good place to start. They were written back in the eighties, and teach more about reading streams and dredging in general. Just because a stream or river has gone dry, doesn't mean the principles in this book doesn't apply. Well worth reading. Couple these books with the placer books available for your state, and you're on your way.

Local Publications

Keep in mind that each state has books on mining. For example, you are interested in a mine or mining district in Nevada. Try this one: "Mining Districts and Mineral Resources of Nevada" by Francis Church Lincoln. This one even has a map, as well as all the historical data on the old mines and what mineral was mined.

There are also books about sections of states and counties, such as "Mines of Eastern Nevada", and "Mines of Storey County". These types of books can narrow a search on a district or mine, and will typically provide more detail than books about statewide mines.

An organized bookshelf is, well, kind of scary.

Chapter Two

Mining Overseas

A strange place for this topic, perhaps. But then again, maybe not. Yeah, there's gold there. Lots of gold. All the way down into South America. Question is, risk versus reward. Even if you are from the country in question, well, think about what you know. Here, in the US, you may have to deal with every government agency up to and including God, but you can do it. Overseas, the obstacles can be staggering, and you can bet that regularly, another surprise will rear its ugly head. You might have to partner with a local attorney, and you can bet your first gold nugget that he won't be on the property swinging a shovel. He will, however, lighten your checkbook every chance he gets. And of course, nepotism is alive and well, be prepared for that. You won't believe who is on your payroll, and they won't be swinging shovels, either.

You probably will have to ship your product to a State owned refiner, and you won't believe what you lost when they finish. And you will pay a percentage for the experience. Everyone gets a cut, you know.

You're just going to throw those gold bars in your luggage, right? You have to declare such things. You won't get far, and you won't get back in the United States without getting caught, either. While you are at it, think about *ASTM, logistics, and the quality* of your supplies.

Mexico And Points South

Think about Mexico. There have been many mines nationalized by the Mexican government, in fact, this was pretty much standard procedure for many, many years. A few guys would get together, open and develop the mine, and start production. The government would then seize the company's funds, and imprison the operators. Large corporations were also subjected to this treatment.

Take a drive around Mexico. There a many, many abandoned operations. There are heap leach pads, carbon in leach plants, flotation mills, and all the equipment, abandoned. The equipment rusts away under the Mexican sun. There is no technical knowledge or financing available to restart the projects. The Mexican government really did kill the goose that laid the golden egg.

"Oh", they'll say, "NAFTA has changed everything!" "Why, you can even own property here!" Hey. NAFTA hasn't changed jack. Period. Business as usual. "La Mordida" (the bite) as in bribery, is truly alive and well. You may want property in some of the resort areas, but out in the mining areas, you'll be glad to get out of there.

The further South you go, the worse it gets. The people are great, it's amazing they survive in the poverty. It might surprise you to find out how much they know about mining. They also know what gold is worth. Do you? Is it worth your life?

Keep in mind a prospecting permit is required in these countries. Think about the knowledge you'll part with to get one. Also think about not getting one, and getting caught. It's going to get real ugly.

Tribal Lands

Native American reservations, or tribal lands, are sovereign nations. If you trespass under any circumstances the consequences can be severe. Many a stray hunter has found out the hard way that they should have paid closer attention to the maps. The seizure of guns, vehicles, and other tales of woe are heard every year.

Mining, or even prospecting on a reservation is not going to happen. Every nook and cranny on the reservation is sacred to one person or another, even if they haven't seen it in four generations. The mere mention of mining will release the wrath of the entire tribe upon you, so don't even go there.

Since they are sovereign nations, think about not being paid for services rendered, and then attempting to collect. There are some tribes that do this as a normal part of business, or they will take their own sweet time paying you.

Think about suing a foreign country to recover a bad debt. You might get paid, but you will go broke to get paid.

Also keep in mind that since they are sovereign nations, any infraction must be dealt with in Federal Court if you are not a tribal member. There are not Federal Courts in every small community. Think about a speeding ticket.

Tribes have their own internal politics, so what may be good today can be prohibited tomorrow. Avoid tribal lands, period.

Medical

Mining overseas anywhere can be a huge risk. Insure yourself and family for emergency extraction and or medical evacuation. You don't have to be a rocket scientist to figure out what can go wrong. Check out the homicides in Mexico, or read about the resurgence of Dengue Fever. Think about malaria, if nothing else. Not just Mexico, anywhere outside of the US. Locate a international travel medical clinic, they will have a list of medications you will need wherever you go. Be sure to have a current record of immunizations with you when you travel. Obviously, a current passport and a work permit or visa are must have items.

People say, "Why, I was in Cancun on vacation, and Mexico was wonderful". Try living with the natives outside a compound. Watch the grinding poverty, dogs so starved their hipbones protrude through the skin. The children aren't in much better shape. Then think about breaking an arm or leg in that environment. Would they re-use a syringe? Is that gold you *might* get worth it?

Violence Overseas

Viva La Revolucion!! Not! Here's more advice. Never, ever discuss politics overseas. No opinions, nothing. This will get you seriously injured or quickly killed. The "Ugly American" is alive and well in every country your author has been to. There are some fantastic prospects overseas, believe me. They just aren't worth dying for. Check the State Department's web site before even thinking of going overseas. You might be surprised at what you find out. Practice your hostage routine. You just might need it. Not surprisingly, an amazing number of Americans claim to be Canadian.

Guns

Got your favorite hand cannon loaded for bear, and packed? *There are laws against firearms everywhere there's gold.* This is a quick way to find out what a foreign prison is like. *You have no rights, whatsoever.* And the State Department may rescue you, but it's going to be a while. They are bureaucrats, your welfare is probably the least of their concerns. Eventually, they'll find out about you, and eventually they might get you out. Maybe.

Yeah, you can hire security, and the minute you produce gold, well, all bets are off. When the choice is your life, or a small gold bar, well, adios, my friend! That small gold bar is more money than a security officer will make in about ten years.

Keeping Your Gold

Your placer or lode mine paid off, and now you're ready to go home with all that shiny gold. What about your lawyer partner? What about the requirement that all gold must be shipped to a state or government owned refiner? By the time everyone finishes slicing up your golden pie, you're probably going to owe them. The refiners are told what numbers to produce, so you're screwed at the onset. Throw the bar in your luggage and run for it? Don't get caught. Metallic gold, soil, and dozens of other commodities require an export permit. You also have to declare value upon arrival in the USA. You will most likely take it in the ass, twice. *Do your research!* Check out the State Department's web page.

Sex and Drugs

If the little woman is at home, what will you do for sex? Like to burn a little weed at the end of the day? Like to pack your face with a little blow now and then? You might want to figure out what's going to happen to you. Maybe you should just stay home. Avoid purchasing sex and drugs. You can really get in trouble quick. Here's hoping you don't go there, the inside of the prisons are real hell holes compared to ours. Plus your package is going to fall off. STD's are alive, well and common overseas.

Oops! Killed or injured someone in a car accident? In some countries you go to prison until you pay the price decided by a judge you will probably never see, whether you are right or wrong. After all, you're the one with the money. Just pray that you aren't injured. You need to know what you are getting into.

If you live overseas, you are a couple steps ahead of the game. If you aren't a citizen, well, let the games begin. Either way, good luck, you're going to need it.

Chapter Three

Professional Help

The point of this chapter is that there is professional help out there. Don't be bashful, none of us knows everything we need to know to complete the projects in this book. So, if you find yourself with an issue, well, dig out the phone book, jump on the web or whatever you need to do to find the appropriate help. Trade publications are also a good place to start. Try the ICMJ (International California Mining Journal), and try to find someone in your area to save travel expenses.

There is an old adage, "It's easier to ask for forgiveness than to ask for permission". Not true when you're dealing with any bureaucracy, be it local, State or Federal. Don't be made an example of, get the right information before you start.

Mining Engineer

These guys are great, and tend to have a wide range of knowledge, regardless of your type of operation. They are always current on mining, and mining law, as well as the engineering side. You may not be trying to figure out how to square set timber in a tunnel, but the engineer can help you out. If nothing, a simple consultation about the overview of your project will be useful. You will be pleasantly surprised.

Geologist

These guys are the rock type guys. If you're having trouble with a geological report, or need more, or current information, go to a geologist. In active mining districts, you'll find most geologists work directly for a mine or mining operation. Some will be associated with engineering firms, so if you find an engineer, he will most likely be able to refer you to a geologist, if needed.

Mining Chemist

Or chemical engineer. These guys typically work for a consulting firm, and seem to be about everywhere. Keep in mind that mining chemistry is quite a bit different, and is considered inorganic chemistry. A biologist won't help you, so check around active mining areas, and again, trade publications. Your assayer may be able to help, a lot of lab managers are chemical engineers, or know them. Ask.

Assay Lab

They say 50% of the labs out there are incompetent or crooks. Go to this site on the web: www.ntc.blm.gov/krc/uploads/318/BLMassaylabsreport.pdf

This is a downloadable PDF file. Download the report. Read it. Study it. Understand it. Actually, 50% is low, there are some real idiots out there, and some obvious con artists, in your author's opinion. Check the results on those blank samples as you review the report.

What happened was that the Bureau of Land Management and Arizona Bureau of Mines got together, finally, and prepared *known ore samples and blanks* (no gold) and sent them to about every assay lab in the country. The results would be funny, except that your author and his dad sent samples to a half dozen of these labs back in the day. It won't take you long to figure out *where not* to send your samples. The report also makes a good case for having your own lab, and that's what happened to your author. There are a few good ones still out there, so don't give up hope.

Amazingly, some of the worst of the offenders are Certified Assayers in Arizona. It's a shame this wasn't done long, long ago. The major mines constantly "round robin" samples to each mine lab to check the quality of the work. The last time your author checked, the confidence factor for the "round robin" samples was 99.7%. Typically, commercial labs won't spend the money to have their competition check the quality of their work.

Wet chemical testing can be as important as a fire assay.

Chapter Four

Sampling Your Find

Methods of Sampling

Whatever your type of material is, it is important to obtain representative samples. There are auto samplers for rock samples and mill circuits. The auto samplers will give you an average of the ore pile or whatever you were sampling for the duration of the time the auto sampler was operating.

On the other hand, a sample taken from the circuit at one specific interval gives you a "snapshot" of the values at the time the sample was taken. A good example of this would be the sampling of a CIL (Carbon In Leach) or CIP (Carbon In Pulp) type circuit where a liter of slurry is taken and analyzed for activated carbon content, cyanide strength, or whatever measurement you were after.

A sample from an auto sampler, for example, will show a trend in the circuit for the duration of the sample taken. The circuit can be failing, and the trend may show, but the exact point of time of the failure would be unknown.

A sample taken at any given interval will tell you if the mill circuit is operating normally, or if the process has failed at the time the sample is taken. Some operators refer to these type of samples as "grab" samples, and they are usually taken at regular intervals throughout the shift.

One of the most common errors by the novice is the "one rock" assay. The novice will find a "pretty" rock, and for some reason, have that one rock assayed. The rock will bring in a good assay, and suddenly, the sugar plum fairies abound, and our novice is going to be rich! Our novice has found Gold!

There are tales of woe about these novices borrowing large sums of money, and then are unable to locate more of those rocks, and even where the rock came from for sure. Incredibly, this is a common story.

The Hard Rock Ore Pile

The first order of business is to establish what values you have. This is done by obtaining a representative sample of what you have, in the most scientific, unbiased method that you can. If your blast hole driller or miner was competent, samples of the drill cuttings were saved from each round. Presumably, these samples were assayed to provide direction to the driller and or miner. If so, you should have compiled the results, and have a reliable "head ore" value. If not, well, read on.

Never "grab sample" an ore pile. This is a mistake because you will, unconsciously or not, grab the best samples. This unintentional "high grading" will really hurt later when your mill recoveries are way too low, because the sampling wasn't accurate.

If you can afford the expense, there are a variety of auto samplers out there. They come in different configurations, and may save a lot of work. Check around, see what's out there.

Equipment dealers can usually help in locating one.

If you read Sir T. K. Rose's "Metallurgy of Gold" you will find a description of "coning". This is about the only economic, low tech method of splitting down the sample of your ore pile.

An ore pile of any large size will require an excavator or backhoe of the extendable hoe variety. Typically, the larger the rock size, the larger the required sample will be. Take out the tape measure, measure the pile, and draw a grid over a sketch of the ore pile. Suppose, for instance, your ore pile is 50' feet wide, 100' feet long by 10' feet high. That would be 50,000 cubic feet or 1851 cubic yards. Hard rock typically runs 3,000 to 3,500 pounds per yard, so roughly you would have 27,765 to 32,392 tons of ore. This doesn't account for moisture, by the way. It may be easier to mark your grid with highway paint.

Take your tractor, and cut the trenches on your grid. Pile the ore, flatten the pile by one third or so of the height. Take some highway paint, and divide the pile into quarters. Remove two of the quarters, pile the quarters into another pile, and quarter again. Eventually, you will have quartered until the sample size is what you need for assays, leach tests, or whatever.

It is always helpful to crush the original cone to minus (less than) one inch in diameter. This will produce much more accurate results. If you had an astute miner, hope he shot the rounds hard enough to produce mine run ore of about minus three inches in diameter.

In the ore pile described above you would want to wind up with about two tons. Cone again, until you have a two 55 gallon samples. Seal one sample in a drum with an airtight lid. Archive the sample. This is insurance in case you have to re-test the ore. This can be very handy to have in the event of any legal challenges, or need to re-test the ore pile.

Now is the time to crush the rest of the sample to a ½ or ¾ minus (less than ½ or ¾ inch). A 6" X 6" lab jaw crusher will be handy for this. Find a Jones splitter (riffle splitter) and split the sample down to say, five gallons. Split that down to say, two pounds. Now you're ready to head for the assay lab and get some numbers. Your assayer may be able to help with crushing, splitting and sampling.

The Dump Stock

Back in the day, mines shipped ore by way of truck, rail, or maybe even wagon. The stock could be fifty tons, or 50,000 tons or even more, depending on the size of the mine. There were times the conveyance simply couldn't come to the mine, be it inclement weather, bankruptcy, floods, snow or any other reason. Rather than shut the mine down, the managers or owners would simply stockpile the mined ore in a pile off to one side or another. The waste or "gangue" went to another pile, commonly called "tails".

Most of these mines were shaft operations, and the operator would run a set of tracks for the ore carts to a suitable location for the dump. To the novice, all those piles look like tail dumps, but not all of them were.

Then along came World War II, and that ended a lot of mines literally overnight. The stockpiles of ore were left sitting, and some still are. They are found from time to time by conscientious prospectors willing to put the effort into locating them.
Your author has found them by looking closely at the remains of the mine, and figuring out what the piles of rock were. Then a representative sample was assayed, and with suitable

results, negotiations were started to purchase the dump stocks, or if abandoned, new claims were filed to take ownership of the property where the dump stocks were located.

After legally securing the rights to the ore, the research begins to determine the best method to recover the values.

Essentially the same process would be followed as with sampling the ore pile, with a couple of notes. First, understand that if the ore has been on the surface for many years, the outside of the pile has oxidized and weathered from the elements. Scrape off the top foot or so, and then take your sample. The chemistry for oxides as opposed to sulfides will be completely different, so don't skew your sample. Get that oxidized ore scraped off.

Usually, there is a noticeable difference in the ore that you can see by just grabbing a shovel and digging into the pile.

Also note that your dump stock is most likely built into the side of a hill or mountainside. Don't dig into the hillside or mountain side, you will dilute your values with the waste under or behind the dump. Be sure to have a suitable area scraped clean to do the coning as previously described.

Usually, the old timers shot their rounds hard, and the material should be smaller in diameter than waste, but if not, get a crusher on site, or transport the material to a more feasible location for coning. A concrete pad can be a wonderful thing for scooping the ore around.

Don't forget to archive your sample as described above, get that sample ready, and then you are off to the assay lab.

Old Mine Dumps

Old mine dumps can be viable projects as well, and there are thousands of them out there. Think about the historic prices of gold. In 1792, gold was $19.75 per troy ounce. In 1834, it was raised to $20.67 per troy ounce, and in 1934 gold was raised to $35.00 per troy ounce. The point is that the old timers weren't after low grade ore. So what were they after? The good high grade free milling gold in quartz was the most sought after ore, and many deposits of complex ores were walked across because the technology to process the ores and have a reasonable recovery and profit simply didn't exist. Another factor one has to think about is inflation. From 1913 to 2014, inflation was 2,302.5%. An item that cost $1.00 in 1913 would cost $24.03 in 2014. So equipment cost more, consumables cost more, and of course, if an ore had a low profit margin, it was simply skipped over.

If you are looking for old dumps, stay away from shafts, and keep out of tunnels. Shafts erode around and underneath the collar, and will cave away, dropping you into the shaft. If you are buried, you will most likely stay there, since the risk of recovering your body would be too great for rescuers. Some tunnels have a winze inside. A winze is a shaft within a tunnel. Usually, the old timers put down a few two or three inch planks, and skipped across the shaft. Now, those planks are dry rotted, and you will plunge into the shaft as previously described.

Sample the old dumps the same as dump stocks. Also remember the original ore mined will be at the very bottom or back of the dump.

What was not economical then could certainly be economical now, with gold prices hovering around $1,300.00 per troy ounce. Perhaps you could re-claim the old dump, and bring in your portable mill and process it on the spot. A lot of old mines "made", or accumulated water, so keep that in mind. You may be able to access the water.

The Placer Claim

Placer mining is far and away the least costly of any mining process out there. In fact, in some situations, placer equipment can be used to recover values from old mill tailings and whatnot. Remember, if it will pan, it will concentrate to a greater or lesser degree.

Placer sampling is a bit different. Get yourself a topographical map of your claim(s) and copy the map. Draw the grid on the copy. The big difference here is you will need to trench all the way across the claim, if you can. If not, dig a test pit on each point on the grid. The trenches are far and away the most representative of the claim. The other thing to remember is to fill in the holes and trenches before you leave. You can unintentionally trap wildlife, livestock, or some kid on an ATV if you don't. Remember that a mining claim entitles you to mineral rights only. If you are on a patented claim, that is private property, and you can restrict access by the public to the property. Also note that the operation of heavy equipment is allowed if the zoning is appropriate. Check the zoning on your property.

If you are on a non-patented mining claim, surface use (cattle grazing, tourists camping, etc.) will continue until you begin an active mine, at which time you will legally be able to restrict public access for safety reasons. This must be part of the surface disturbance portion of your permit, or mining plan. The plan must be submitted to the Bureau of Land Management, Forest Service, your State or applicable managing agency.

You should also cross claim your placer claims with lode claims. If you don't someone can come on your placer claims, and start a hard rock mine. Who needs an open pit in the middle of their placer claims? Think about that, it is a common preventative practice.

Back in the day, placers were sampled with a bucket auger. The auger had a clamshell around it, and the entire sample was removed from the hole. Technically, this is the best way to sample, however, that style of powered bucket augers have gone the way of the dinosaur.

Peruse the web, see what's out there. How deep do you have to dig to hit bedrock? It might be easier to go with a backhoe, or push off some overburden with your handy D-10.

The best way to process your samples is to get a pilot mill on the claim, where there is water available. Remember that any usage of surface water must be part of the permitting process. Pay attention to the information on clear water technology later in this book, it will enhance your recoveries.

We will assume you are not able to find a small plant, and get permitted to run on your claims, and your samples were put in barrels with a lid, and transported elsewhere to be tested in a pilot mill.

Go to Chapter Nineteen, Pilot Testing, and read "The Placer Claim".

Old Mill Tailings

These old tailings are great to work with for one major reason. All the crushing and grinding has already been done. The best tailings, or highest in value that your author has seen over the years were from flotation mills. Don't overlook cyanide tails, they can have recoverable values as well.

Flotation mills (or just float mills) were normally used in ores with a high sulfide content, or some sort of offending penalty element present, such as arsenic. The chemistry could be altered to float or depress most any element.

Typical reagents were pine oil, diesel fuel, a frother, and usually a xanthate of one sort or another. Potassium Amyl Xanthate was popular until recently. You could tell a float operation by the stench as you approached.

American Cyanamid invented and sold these reagents for many years. They worked, and worked well. They also had a couple of books, Mining Chemical Handbooks, or Mineral Dressing Notes, that covered every aspect of flotation. There are chapters on promoters, (sulfide, metallic, non-metallic and metallic oxide) dewatering aids, frothers, modifying agents, and flocculents. They also list recipes by ore type. If you ever find these books, grab them. The last update was in 1986, they are great reference material.

Flotation is basically the process of making a select particle of mineral hydrophobic through selective chemistry. The frother picked up the particle, and carried it to a trough for recovery. It took a lot of test work on the particle size and the chemistry to get there, but the process worked quite well on some ores. Keep in mind any non-metallic particles or water soluble mineral went to the tailings pond for your future mining enjoyment. Think about any of the platinum salts, gold chloride, silver chloride, and so forth. Some metallic particles went to tail, as well.

Sampling is a lot like the sampling of a placer. Get a backhoe with a very narrow bucket, or if you can find one, a bucket auger.

Make sure you have legal rights to the tails, or you'll have ten other people thinking you've found the mother lode, and jumping your property. Also remember that you are assuming the liability if there is something nasty there, such as cadmium or beryllium.

Anyway, do the grid, and pull a sample of say, 35 gallons per hole. If the tails are fairly shallow, you can break the samples into more than one part. Say the first three feet in one barrel, the second three feet in another barrel, and so forth. Once you've sampled what your budget will allow, fill your holes! Cone as previously described, and save a reference sample.

For a quick preview, grab your pan, or find a lab table and gather about 30 grams of concentrate. Have it assayed, and if there are acceptable values, go for a spectrographic analysis, then an SEM Micro probe. Don't forget to assay a head sample for a base line on the values.

Keep in mind that as the mine progressed, the characteristics of the ore changed. If the mine got ahead of the chemist, some pretty good stuff escaped, and is in one of those layers. That happened a lot, and that is what you are looking for. You may be able to scrape off the low grade to get to that high grade layer, if that's how it works out.

Cyanide tails can also carry good values. A 40 micron particle of gold takes 17 days to go in solution in cyanide. What if the grind was off for a few days, letting coarse particles into the circuit? Discharge screens in ball mills fail, so it does happen.

Cyanide tails that have set for decades exposed to the elements tend to oxidize over time, and the sodium hydroxide used to maintain protective alkalinity helps the process. So does the hot desert sun. Cyanide has been in use over a hundred years, so there are some really nice, old tailings out there. Don't breathe the dust. It is a strong alkaline, and will damage your respiratory tract.

Once you're ready to process, the work done so far will help determine the way to process the tailings. You won't have to crush or (hopefully) grind. You will have to put the material through a screen deck to remove tree branches, rocks, and other debris.

At this point, with the information you have accumulated, invest a little time with a good mining engineer. Get a direction, do more homework, check permitting and so forth. Think about access to water!

Cyanide tails in Arizona, completely oxidized by chemistry and weather.

Chapter Five

What Are Your Values?

Now that you have a representative sample, the question is what values you have in your ore pile, dump stock, mine dump, or old tailings.

First and foremost, send a pound to several assay labs for fire assay. Start here, and see what the numbers look like, and see if there is any consistency in the results between the labs you submitted your samples to. Sometimes a fellow miner can recommend a good lab to start with. Keep in mind, not all labs are created equal. How the results are achieved on your samples can vary considerably from lab to lab. There are even so called "labs" that will provide numbers on demand. You tell them what you want, they produce the numbers you ask for, and voila', you're off to do another mining scam. Don't go there. All the western States and most of the others will vigorously prosecute these offenders and probably you as well.

Salting Your Claim

This isn't the "Old West", but salting claims is still a common practice. If you are buying a claim, or ore pile, or whatever, you should be aware of this shadier aspect of mining. The bad guys really are alive and well. Did you get a mental image of some old geezer pouring gold dust in a shotgun shell? Well, it still works as long as the values can be confined to a fairly small area, such as a drill face in a tunnel or shaft. The old geezer would simply fire a few shells in the dig face, and let you "discover" the values.

The point is, the gold won't penetrate into the rock, so break off some of the gold laced rock, and look under or behind the gold. If natural, the gold will keep going into the rock. No gold? Better look real close at what you're doing.

Another popular trick is for our geezer to find some quartz, or other mineral that is fractured. The rock is heated with a torch, and gold solder is applied to the rock, and melts into the fractures. Just remember, if it's natural gold, it will go all the way through the rock. So break the rock. The solder job will be pretty obvious.

Always look for specimens that have been seeded onto the property or rock in question. Look at the matrix the gold is in. Is that matrix present and normal for the property? For instance, you have a free milling gold in a rusty quartz. And suddenly you find a chunk of manganese stained (pale purple) with visible gold. That should be your first clue something is bad wrong. The mismatch of the matrix would be an indicator of "float" or a chunk of ore that has somehow been transported to the current location. Not a good idea to take a surface sample if in doubt. Dig down a foot or two.

Placer? Easy. Sprinkle fine gold particles in a few low pressure areas, and steer the mark (you) to them. Golly, gee, you've found gold!

You should put any gold sample in doubt under a microscope and look for color. Typically, gold in nature is about 80 fine, a pale yellow color. Suddenly, you start finding dark yellow gold, 90 fine or so. Think about what you really should expect to see. Also remember that placer gold is smooth, and rounded. Gold out of the ore matrix should be rough edged, and jagged. Pay attention.

Probably the most evil trick is the use of Atomic Absorption (A-A) standard. It is a pale yellow liquid, gold dissolved in aqua regia, which is diluted to the appropriate PPM for the A-A standard. It is 1,000 PPM off the shelf. If you put this liquid in a spray bottle, and spray select areas with it, the liquid will dry out, and the sample will fire assay beautifully. Don't panic...Split your sample in half, rinse one half in hot water, dry it, and send *both* to the assay lab. If the unwashed sample comes back way high, the other washed sample way low, say more than 20% variation, re-sample, and rinse the whole sample. Re-submit for assay. Think about what someone is trying to do to you.

The A-A standard trick has become a staple the last 20 years or so, especially in chemical leaching. The leach system is set up with "secret" chemistry, the mark (you) take a sample, which basically shows zero on the A-A. The "secret" ingredient is added, and in a short period of time, the mark (you) sample again, and voila', all sorts of gold is in solution. One of their favorite tricks it to have the mark (you) add the "activator" or whatever they call it, and take the samples. Adds credibility to what they are doing, and since you are participating, it's hard to see what's going on. How about if you just get a pound or two of the ore before it gets in the tank? Assay that. It all has to start with the ore, does it not? *Note that this deception will work with any chemistry, not just cyanide.*

The other thing to keep in mind is that standards are available for every element known to man. Even platinum. How about if they salt your spectrograph sample with a spray bottle as above, and let your chemist do their dirty work? You're looking for gold, and you start showing high platinum on your spectrograph sample. Your chemist is basically assisting in the salting without having a clue. Re-sample, rinse well, preferably in hot water, and analyze again. Remember, one assay doesn't make a mine. Be paranoid, and verify everything before investing any money.

And now you know why there are disclaimers on assay reports.

High Grading Your Samples

This is something that is done inadvertently, as opposed to salting. It is, in most cases, a psychological mind set that people get into. Or, it can be done intentionally, with deliberation. That's still salting, folks.

A client had a property in South America, and had the locals, who had no sampling or mining background, sample his property. The numbers came in at a solid 2.0 OPT gold, one right after the other. Based on that information, the client went to the property, and pulled his own samples. Those samples came in at about 0.05 OPT gold. This happened twice, and finally, the client had the property professionally sampled.

Come to find out, the locals, who really wanted the operation to go had intentionally high graded the samples. The client, who had no idea what he was doing, had unintentionally low graded the samples out of ignorance. The professionally collected samples averaged out around 1.00 OPT gold.

Many a miner has kicked a good looking piece of ore into a channel cut that is being sampled, thinking it won't make much of a difference. Not true. When in doubt, it is best to have an uninvolved third party who knows the correct procedure do the sampling. "Grab" samples are notoriously inaccurate for the reasons previously stated.

Remember to always mark your sample locations carefully, so you may return to the exact site, even years later if necessary. Pictures are good, as well.

The "Gold Pan" Assay

The old timers called this "counting colors". Don't waste your time. The reasons this doesn't work are many. The panner, no matter how skilled, never weighs how much he pans. That's a pretty fundamental requirement for a quantitative (how much) assay. Did those colors come from one pound? Ten? Wet or dry weight? There's also that phenomena called "flour gold". It will take thousands of particles to make a pennyweight, and yes, they are visible in the pan. Particle size has everything to do with an assay. These "eyeball" assays never work, and head to head, in a controlled lab environment they will fail every time when compared to a properly done fire assay on a representative sample. Never let wishful thinking be your guide. Get some solid science on your side.

Panning a sample is a good idea in the fact that all the heavier elements will be in the pan and visible. Look for gold, and such things as pyrite in the pan. There are always plenty of black sands or other "grey metal" visible. Don't get excited about the grey metal, there are dozens of minerals that are grey, or gray in nature. They are usually an iron compound, or maybe tin, or molybdenum, not platinum. When you do your spectrographic analysis a little later on, you'll know precisely what that metal is.

In the lab, there is a procedure called vanning, where a large watch glass is used to wash a teaspoon sized sample. The assayer is looking for the brittle metallic elements, such as pyrite, and silica compounds. These minerals in excess can have a profound effect on the fire assay, and obviously will have a bearing on the assay and your milling operation.

Always pan some of your ore. Get a feel for the metallic elements. See any gold, iron pyrite, or excessive quantities of grey metal? Identify these elements before you start. The multi-element spectrograph will help a lot with this.

Dredging the Rio Caldera in Columbia, S. A. The dredge is a Keene triple sluice five inch. The author is in the foreground.

Old cyanide tails near Oatman, AZ. There are many locations like this across the West. Note that the tails are completely oxidized.

Abandoned ball mill, pumps, cyclone and pachuca tank in southern Mexico. The operation was a flotation mill. The sulfides are slowly dissolving the metal.

Abandoned mining operation in Columbia, S.A. Home made ball mills in rear, Denver equipment jaw crusher in foreground.

Chapter Six

The Lab

Do you need a Lab?

Yeah, you do. If you can. The problem with all labs is turnaround time, not to mention incompetence, and out and out fraud. Getting your results two weeks after you submit samples will not work in the typical mining or milling environment. If you can't get a one or two day turnaround from your lab, you need to seriously investigate putting in a lab. All major mines have access to a fire assay lab, an A-A facility, as well as a metallurgical lab. These facilities will be located on the mine site for fast, easy access.

You need to understand that any lab will have drawbacks. A simple fire assay lab will work for 99% of your needs. A fire lab can produce results in as little as one day, and has a of detection limit of .001 OPT gold. Also note that Carlin type ores fire assay quite nicely, contrary to rumors. All of the mines on the Carlin Trend use fire assay.

You would be able to assay rock samples, mill solids, solutions, carbon and other organic solids. There are many assay books available these days, and Action Mining has a "home study" fire assay course. They also have "kits" with the basics. There are no schools that teach fire assay any longer. You should know that to assay for third parties (commercially) requires certification in most mining states. When you think about the liability, you would be insane to assay for the public.

Experienced assayers are out there. A general understanding of basic chemistry is an absolute necessity, as is a working knowledge of lab safety. A lot of experienced fire assayers can also operate an AA.

If you end up running a chemical leach operation, plan on adding an Atomic Absorption Spectrophotometer, or AA to your lab facility. This will enable you to analyze leach solutions onsite with a very rapid turnaround time. Most basic chemist types should be familiar with the operation of an AA, and they are very common in the water treatment industry. AA's normally analyze for one element, and require standards, setup, and another lamp to analyze for another element. The older Instrumentation Labs AA's had a turret that held five lamps, and could be switched rapidly from element to element. If you purchase a used machine, always purchase a machine used for water analysis. There will be little corrosion in the electronics from mineral acids.

The advantages to having a lab are many. The single, main advantage is access to process numbers in a relatively short time. Something as simple as a screen failure can cost you an enormous amount of money. Once your values have gone to the tailings pond or waste pile, it's not feasible to shut down and re-run the tails. Going after the lost values compounds this error, because you have to shut your mill down, and lose production time. All over a lousy screen. Stay informed, get a lab running if you can.

Drawbacks for a Fire Lab

The fire assay relies on PbO, lead monoxide, or litharge, to collect the precious metals in a crucible fusion. Lead contamination is an issue, and all crucibles, cupels, and slag are hazardous waste, and must be collected and disposed of according to current environmental law. There are companies, such as Laidlaw that specialize in the disposal of lead waste.

Lead contamination is an issue for employees, as well as the scrubber or bag house that ventilates the fume hood. At assay temperatures, during both fusion and cupellation, metal oxides evolve and must be vented away from the operator(s). Other toxic and some poisonous metal oxides are also evolved as the sample being assayed is heated. Adequate forced ventilation is a must. As the metal oxides precipitate in the ventilation system, they become a very fine powder known as "flue dust", and the discharge of this dust into the atmosphere is against regulations. Some metal oxides do not cool enough to precipitate, and are visible as smoke. This smoke can't be discharged, either. Hence the bag house or scrubber.

Obviously, this is an industrial process, and requires the proper zoning. If your mill is the hills somewhere, check with the county, BLM, or appropriate agency as you are permitting your milling operation. Most states and counties will require permitting, or the initial plan must contain plans for the assay lab, and it will be part of the project.

Plan on providing dust and mist respirators and coveralls for your employees, and if possible, on site showers for the end of shift is a good idea. The lead contamination on street clothing and shoes will go home with the employee and contaminate carpet, children, and pets. Heavy, leather slip on steel toe footwear is a good thing. Think about unlacing a calf high work boot as 2000° slag is cooling on your foot.

A local industrial medical clinic will be handy, if you have one. If not, any MD can do blood draws. All employees should have a blood test for lead and mercury before working. Cover your ass! Establish a base line for each employee and yourself before you start. There are lots of sources of lead out there, and someone you hire may already have an elevated blood lead level. Then you assume the liability of their medical treatment. Re-test for lead and mercury quarterly. If an employee has rising blood lead levels, poor hygiene is usually the cause. Respirators must be worn in the fire room, period. Employees must wash their hands thoroughly before eating or smoking.

Drawbacks for an AA Operation

AA's need to be in their own room, and the room must be kept scrupulously clean. Dust will contaminate the whole process. AA's require a set of standards for each element being analyzed. The standards rarely last more than a month, and must be replaced constantly. Standards are made in volumetric flasks of various sizes, and volumetric flasks aren't cheap. The standards required and the concentration of the standards is predicated on the elements you will be analyzing for. Keep the caps on the flasks to prevent contamination. The room must be ventilated, preferably with a small hood over the burner.

The solutions you will be analyzing will be either acid or base (alkaline) and you certainly won't want to breathe the vapor.

The single most important thing to know is *the rule of identical matrices*. Don't build standards on DI water. Calibrate the machine with the same solution you will be analyzing. For example, if you are running a 2 lb per ton cyanide solution with protective alkalinity (pH) set with sodium hydroxide in your mill circuit, that same solution is what your standards should be made with. Also, when you have finished analyzing samples, let the machine aspirate DI water for at least a minute to flush the alkali or acid solution out of the burner head.

Note that AA's require compressed air and acetylene to create the flame at the burner. Make sure you have a moisture and oil filter inline on the air line. Some elements may require nitrous oxide. Both acetylene and nitrous oxide are available at any local welding supply house. Nitrous oxide requires a different pressure regulator than acetylene.

AA's are fairly common, and also fairly inexpensive used. They can be expensive new. Remember to buy a water analysis machine if you buy used. You should be able to get a warranty from most dealers. Perkin Elmer, Varian, Instrumentation Labs, and other brands are out there.

AA's require service at manufacturer's recommended intervals. The mirrors need cleaned, among other things. Poor ventilation will cause the mirrors to fog with residue, so make sure to ventilate adequately. As the mirrors fog, sensitivity decreases.

When new, AA's come with good manuals. Make sure your machine comes with one, new or used. Volumetric flasks, standards, lamps and other can be purchased at most chemical supply houses. Check out Fisher Scientific, Van Waters and Rogers, and search the web. Used glassware is acceptable if used in water analysis. Don't buy used glassware from mining operations or high schools, that glassware is rarely properly cleaned. It can contaminate your work.

ICP and DCP

Inductively Coupled Plasma and Directly Coupled Plasma machines are a considerable step up in the analysis of solutions, and are considerably more expensive than an AA. They are capable of higher temperatures at the burner head, and are computer controlled. The learning curve is steeper than the AA by far.

In the mining industry, DCP is less desirable due to the tendency of soluble silicates to plug the orifices on the burner head.

These are the machines that give you a multi-element analysis. A fairly large sample is digested, diluted to volume, and as the solution is aspirated, the machine scans for multiple elements. Nice to have, expensive to have, but not used daily, so not a need to have item at a typical small or medium mine site.

Major mines moving a million tons or so a day will have a ICP or DCP on site, and probably a Scanning Electron Microscope as well. Moving massive tonnages through an ore body requires a lot of answers fast, and this is how it's done. They can't wait two weeks for an out of state lab to provide answers. They will also have extensive fire assay capabilities, turning as many as 1,000 fire assays per shift. On site turnaround is 24 hours. Solutions are usually analyzed and reported on the same shift as the sample was taken.

X-Ray Fluorescence

This method of analysis becoming more and more popular in the mining industry, especially for loaded activated carbon. It is not as accurate as a fire assay, with somewhat lower detection limits, but the rapid analysis offsets the detection limits. The latest generation of XRF equipment is smaller, and is actually portable. Search the web, take a look at what's new. XRF equipment is available now that can be used for bore hole analysis, and more.

Fire Assay

For nearly a century, banks would loan money on a fire assay only. The results of the fire assay were what the bank considered recoverable gold. The fire assay is also both quantitative (how much) and qualitative (what it is). The simple fact that only precious metals survive cupellation, and the subsequent parting of the silver from the gold are what makes the fire assay the fastest, cheapest, most accurate method around.

Fire assay is the oldest known method of analysis of precious metals, dating back to ancient Troy, around 2600 B.C. There are references in the Bible that date back to 1300 B.C. There are published fire assay texts that date back to 900 A.D. So yeah, fire assay has been around a while. The process has evolved, however.

What will the lab do with your ore? First, it will be dried at 250° F, then crushed to ½ inch minus, then pulverized to at least -100 mesh. After being blended and split, a portion of an assay ton will be weighed into a refractory crucible. A flux, consisting of litharge, silica, soda ash and borax glass will be added and mixed. If your sample is an oxide, a reducing agent such as common flour is added as well. A silver inquart will be added to help collect the gold. Silver has an affinity for gold, and lead has an affinity for silver. The crucible and contents are heated to 1900° F or thereabouts for approximately an hour.

The litharge will have been reduced to elemental lead, which will have collected the silver, which in turn, will have collected the gold. The entire molten contents are poured into a pouring mold, and the lead will settle to the bottom, the slag, which is the waste portion of the sample will be above the lead. The litharge at assay temperatures decomposes the gangue, or waste portion of the sample to slag, or a borosilicate glass.

When the pour has cooled, the lead portion is recovered as a button, and placed in a magnesite or bone ash cupel. A cupel has a hemispherical depression in the top that holds the lead button. The cupel is heated to approximately 1850° F. The cupel absorbs most of the lead. What remains is a tiny bead of all the precious metal in the sample. The bead is digested, weighed and the gold and silver values are calculated.

A skilled assayer can assay for platinum, as well. The detection limit of a fire assay is 0.001 ounces per ton (OPT) gold. Your assay should be reported in ounces per ton, however modern chemistry stresses parts per million (PPM) since the assay ton system is considered archaic. If your assay is reported in parts per million, convert the assay as follows: 1 PPM = .029166 Troy ounces per ton. If you have 10 PPM, you have .291 OPT.

Note that gold is reported to three decimal places, and silver to two decimal places. If your report says TR or NG, TR means trace, or present but not weighable, and NG means "no gold".

Always ask for your sample "reject" back. This is what is left after your sample has prepared, and what is left after the assay. Label it as reject, and keep it for use as a check sample. Send it to other labs to check their numbers, or use it as a control when you send in another set of samples to the same lab.

Another thought to keep in mind is that when you feel comfortable with a lab and the results reported, talk with the manager. Most labs have a quantity discount, and once they know you, perhaps you can get expedited results or other benefits. Always keep your bill current, the assay lab will come to know you, and show their appreciation.

With any assay lab, always check their work. Go to the local gravel pit, or an old local mine and get a bucket of rock. You can even go to the local building supply and get some gravel (not cement) to make cement with. The assumption would be that there is no, or very low gold or silver in this material. Use this for your check assays as well. If the assay comes back high on that old dump rock, head out again for another sample, maybe you've discovered the Mother Lode!

At the time this was written, a standard fire assay was anywhere from $25 to $50 per sample. That is still a bargain, considering the value of the information you receive, and the most fundamental information you will need. Don't skimp on the analytical work. Remember that major mines will be doing 500 to 1,000 fire assays per day. You will also find that every successful mine has an assay lab on site. They know what their ore body and mill is doing at all times by analyzing samples from blast holes and the process circuit.

There are those out there that will tell you that the microscopic "Carlin Trend" ores will not fire assay. Strange that both Newmont and Barrick Goldstrike, not to mention smaller operations on the Carlin trend all utilize the fire assay. That rumor was probably started by the labs that don't have fire assay capability. The facts are the facts, and if you go to the major mine's web site, you will see that they are primarily using fire assay. Instrumental methods may supplement the fire assay, such as an "AA" finish on the parted bead.

At the bottom, or somewhere on the report, you will notice a disclaimer. The disclaimer will limit the liability for the assayer or lab to the price of the assay. The disclaimer should state that the sample must be taken by lab personnel for accuracy and a guarantee of the results. This is because most people do not understand the concept of a representative sample. The disclaimer also recognizes that people will high grade samples, whether intentional or not.

From this point on, *for the sake of discussion*, we will assume you have a head grade of 5 OPT gold, and 2 OPT silver. See the assay report on the next page. _Actually, this is real high grade, and you are not likely to see assay numbers like this, but for our purposes, as a demonstration, this will work._

The Assay Report:

<table>
<tr><td colspan="5" align="center">

Acme Assay Company
123 Golden Way
Somewhere, NV 89706

Assay Report - Gold & Silver

Client: John Smith Mine   123 Easy Way Rd.   Gold Mine, AZ

</td></tr>
</table>

Sample ID:	Sample Name	Au, Oz/Ton	Ag, Oz/Ton
1.	1' to 5'	4.990	1.89
2.	6' to 10'	4.990	1.90
3.	11' to 15'	5.030	1.87
4.	16' to 20'	4.980	1.99
5.	21' to 25'	5.130	2.20
6.	26' to 30'	5.001	2.30
7.			
8.			
9.			
10.		(Averages:)	
		5.020	2.03

Assayer: Leroy Jones
Date: 2/14/2015

Well, finally! Our assay report is back. Above is a typical report, this one has both gold and silver reported in Ounces Per Ton (OPT). A ton would be 2,000 lbs in this case. Note that the values are distributed evenly. If you had not mixed your sample well, or grab sampled, you would probably have a high flyer or two, and probably a low number or two. When in doubt, re-sample, and re-assay. At the very least, re-submit the flyers, both high and low, and see if that helps. Note that the averages are not normally provided by labs, they were provided for clarification.

At this point you haven't invested a huge amount of money in equipment and such, so if in doubt, do it again and make sure you are where you want to be.

Screen Fractions Analysis

At this point, we know what you have, the next question is, where, exactly, in the mineral are the values?

Here's a quick explanation of process or milling terminology. If a particle is small enough to pass through a certain size screen opening, it is expressed as "minus" with the minus sign, then the mesh size opening in the screen. Such as -100 mesh. The particle size can also be expressed as "plus", or larger than 100 mesh. A particle that is -100 mesh, +120 mesh will pass through a 100 mesh screen, but not a 120 mesh screen. This method of describing particle size is commonly used every day in mining. The period at the end of this sentence is about 80 mesh. Talcum powder is approximately -200 mesh.

To further confuse you, there are two types of screen sizes in use. The ASTM sizes, and the Tyler sizes. Tyler is the most common. If you look on the side of any test sieve, the label will tell you the opening size in the mesh and in thousandths of an inch, as well as which type of screen it is. Don't mix brands of sieves.

The next set of numbers you need is called a screen fractions analysis. This information will provide the information necessary for a particle size distribution curve. The idea is to see where the gold is in each screen fraction of your ore. This information will tell you how fine you have to grind (mill) your ore to liberate the element you wish to recover. In this case, it is the 5.000 OPT gold (remember we report gold to three decimals) and 2.00 OPT silver (we report silver to two decimals) from the assay earlier in this document.

Most assayers are familiar with this process, and will tell you how much sample is required to perform the analysis. Usually, two to five pounds will be required.

To do this, first the sample will be dried, and crushed. Then, when pulverizing the sample, the operator will crack the plates to about an 1/8th inch, and pulverize the entire sample. The sample is then placed in a top sieve of a stack of sieves on a sieve shaker. The sieves become progressively smaller toward the bottom of the stack, and end with a pan under the last sieve. The sieves are shaken, and each screen fraction over -100 is re-pulverized and re-screened for the fire assay.

In some cases, more sample is required to get enough sample for a fraction of the smaller mesh sizes, but two to five pounds is usually adequate. The sieve sizes can vary, but a normal stack would start with, say -12, then -20, -30, -40, -60, -80, -100, -120, 150, -200, and -400.

Once each fraction is assayed, you can look at the assay report, and literally see where the gold is in each fraction. As an example, if 95% of your values are +100, then a grind of + 80 (-100) should give you access to 95% or more of your values. Grinding further is pointless, and a waste of money.

A huge amount of money is spent over-grinding ores. Why would you grind past the point of mineral liberation? This test is pretty straightforward, yet most small operators aren't aware of the test, don't want to spend the money, or presume to fly by the seat of their pants. You have to know where the values are in relation to particle size to make a decision on the mill process.

If you are lucky, and have visible particles of gold, this ore is "free milling". Again, we would grind to the point of mineral liberation. The idea is to have reasonably close particle sizes in any gravimetric or wet concentration method you may have to use, such as a concentrating table, hydrofuge, bowl, or Neffco bowl. If you overwhelm your table, or whatever method you use, large particles will displace smaller particles, including particles of gold. The smaller particles of gold will go to tails, and your recoveries will be unacceptable.

Another by product of over grinding ores is the creation of -400 mesh ore, which will be what is called "slimes". If you wind up using any kind of wet milling circuit, and you probably will, slimes will be a serious and expensive problem. Slimes are the fine particles that remain suspended in solution that will not settle. Slimes require chemical treatment for settling. Slimes will be discussed in greater detail in the section on water.

If your ore has large, visible gold particles, then you would remove the large particles first. Grind to the size of the largest particles, and remove them by some method of concentration. The ground material is classified (put through a screen) to eliminate the fine material that is smaller than the large particles. For lab purposes, a gold pan can be used, assuming the pan operator is skilled. The advantage is that the large particles are actually visible in the pan. The visible gold is removed. The lighter waste material is dried and re-ground, and if gold is visible, it is panned again. And again, if necessary.

The idea is to remove all the free gold to eliminate the "nugget effect". The remaining fines from the pan or table are assayed to determine whether or not there are values worth pursuing. If you know the sample size you started with, and put the gold through an assay, you can calculate the actual value of the ore.

As an example, if you started with two pounds of ore, that would be 1000th of a ton. After assaying, weigh the recovered gold in grams. The assay step is important because it will eliminate the base elements that can affect the weight of the gold, such as iron. If you recovered 1.5 grams of gold, then you would multiply the weight of the gold by 1000. The result, 1500 grams, divided by 31.1 grams, would be 48.23 troy ounces of precious metal per ton. Have your assayer do a bullion analysis or part a known amount of your precious metal. Suppose your precious metal is 80 fine, or 80 % gold. Take 80% of 48.23 from your previous calculation, and you will see that 38.58 ounces of your metal is fine (pure) gold. It would be a safe assumption that the rest (9.65 ounces) are mostly silver.

There is an old adage about the larger the assay, the more accurate the assay. If you have large particles of free gold, another method is to simply process 500 pounds, or a ton, and actually have a recovery to calculate as previously indicated. Keep in mind that multiple grinds to smaller sizes and the appropriate classification will be necessary to complete the recovery. At the end of the circuit, the tail assay will tell you the efficiency of the circuit. Keep in mind that .08 OPT gold is considered acceptable leach grade ore. There will be discussions about simple gravimetric circuits later on in the book.

At this point, you know the head grade of your ore as well as what the particle size the gold in the ore is. You know what it is, how much there is, and where it is. All of this information will come together to dictate the process you will need to complete an acceptable recovery of the values.

Aha! Our screen fractions assays are back! Study the following report. Note that TR is trace, and NG is no gold. A .001 number is a particle of gold big enough for your assayer to weigh, but obviously is not a value you can recover. The main point is that there really isn't anything worth recovering beyond +100 mesh. The total available in all fractions is 5.113 OPT gold, to +100 is 4.939 OPT, which is 96.6% of the gold available. You could go for the +110, that would give you 98.1%. The question you have to ask is what the cost of the extra grind down to -100 to +110. If necessary, have your assayer do a Bond Work Index on a head sample. This will tell you what it will cost to mill the ore. All you need to know to calculate the Index is the cost of your electricity. Or if you have diesel power, fuel and maintenance costs. The other good news is that gravimetric equipment will work on this ore, and you shouldn't need chemistry to do the job.

Acme Assay Company
123 Golden Way
Somewhere, NV 89706

Assay Report - Gold Only - Screen Fractions #1

John Smith Mine   123 Easy Way Rd.   Gold Mine, AZ

Sample ID:	Sample Name	Au, Oz/Ton	Ag, Oz/Ton
1	-10 to +12	0.001	N/A
2	-12 to +15	TR	N/A
3	-15 to +20	0.002	N/A
4	-20 to +30	0.001	N/A
5	-30 to +40	0.089	N/A
6	-40 to +50	0.150	N/A
7	-50 to +60	0.250	N/A
8	-60to +70	1.400	N/A
9	-70 to +80	1.350	N/A
10	-80 to +90	1.580	N/A
11	-90 to +100	0.200	N/A
12	-100 to +110	0.08	N/A
13	-110 to -120	0.002	N/A
14	-120 to +130	0.001	N/A
15	-130 to +140	0.001	N/A
16	-140 to +150	0.002	N/A
17	-150 to +160	0.001	N/A
18	-160 to +170	TR	N/A
19	-170 to +180	0.001	N/A
20	-180 to +190	TR	N/A
21	-190 to +200	0.001	N/A
22	-200 to -250	0.001	N/A
23	-250 to +300	TR	N/A
24	-300 to +350	TR	N/A
25	-350 to -400	NG	N/A
26	-400	NG	N/A
Note:	NG = No Gold	TR = Trace	

Date: 2/14/2014   Assayer: Leroy Jones

Let's look at another scenario on Screen Fractions #2 below. We have a total of 5.113 OPT gold available, but look where it is. The real values start at -190, and go clear through -400. In fact, 99.5% of the gold follows in this range. Better start permitting for some chemistry.

Acme Assay Company
23 Golden Way
Somewhere, NV 89706

Assay Report - Gold Only - Screen Fractions #2

John Smith Mine   **123 Easy Way Rd.**   **Gold Mine, AZ**

Sample ID:	Sample Name	Au, Oz/Ton	Ag, Oz/Ton
1	-10 to +12	TR	N/A
2	-12 to +15	TR	N/A
3	-15 to +20	0.001	N/A
4	-20 to +30	0.001	N/A
5	-30 to +40	0.001	N/A
6	-40 to +50	TR	N/A
7	-50 to +60	TR	N/A
8	-60to +70	0.001	N/A
9	-70 to +80	0.002	N/A
10	-80 to +90	0.001	N/A
11	-90 to +100	0.001	N/A
12	-100 to +110	0.001	N/A
13	-110 to -120	0.001	N/A
14	-120 to +130	0.001	N/A
15	-130 to +140	0.001	N/A
16	-140 to +150	0.002	N/A
17	-150 to +160	0.001	N/A
18	-160 to +170	0.005	N/A
19	-170 to +180	0.006	N/A
20	-180 to +190	0.002	N/A
21	-190 to +200	0.060	N/A
22	-200 to -250	0.890	N/A
23	-250 to +300	1.308	N/A
24	-300 to +350	1.219	N/A
25	-350 to -400	0.519	N/A
26	-400	1.089	N/A
Note:	NG = No Gold TR = Trace		

Date: 2/14/2014   Assayer: Leroy Jones

There are a lot of people in the equipment business that will swear they can recover 90% plus of -400 gold particles using some gizmo. That's bull. The vast majority of all ores have a clay content that can be from a small percentage to 10 or 20 percent. That's one variable of many that will stop any kind of decent recovery.

If your gold particles are flat, it's really going to get interesting, because the principles of laminar flow are going to kick in, and the clay will suspend the gold particles right out the end of the machine. Oops! What? No sale? Remember, most dealers could care less about anything but your checkbook. If the equipment dealer can depress the fine gold, and provide a good recovery, is he going to tell you how he did it?

Sodium hydroxide is a depressant for fine gold, so dip a strip of pH paper in the water. If the pH is high you might ask, see if they will at least tell you the truth. Feel some of the water between you fingers. Is it slippery? If so, that's sodium hydroxide, also known as caustic soda, or Red Devil Lye.

Don't watch what they have on hand run on their equipment, take your ground ore, check their recovery with your ore. A five gallon bucket will do for starters. Grab your entire sample back when they are done, and get it to your assayer for a head and tail assay. The dealer will have the equipment set for a certain type of ore for demonstration purposes. If he tells you no adjustment is required on the equipment, well, watch out. That's not very likely, so verify the results.

And yeah, there are idiots out there that are still using mercury. If so, run, don't walk off the property. Don't forget your bucket! Don't play with mercury, it's illegal.

Next, have your assayer do a hot cyanide shake and assay the solution for gold. See what that recovery looks like.

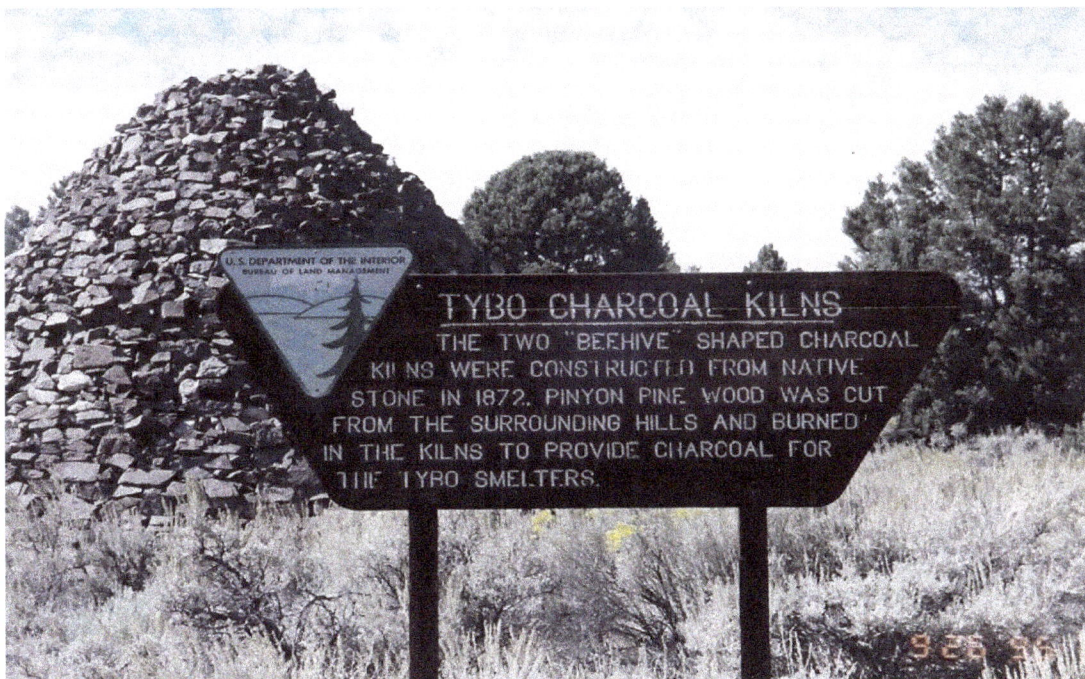

Abandoned charcoal kilns near the ghost town of Tybo, NV.

Chapter Seven

Learning To Identify Your Ore

Ore Types

Now we know what's in the ore, but what is it? In case you wonder, some types of ores are very difficult to work with, namely refractory or complex ores. What do you have? An oxide, sulfide, chloride, or telluride? The most common ores today are sulfides, and a pre-treatment or two is required to mill, or process this type of ore. If you have pyrite, or iron pyrite, your ore is a sulfide. Pyrite would be FeS_2. Pyrite is a brassy, brittle shiny metal with a square, or cubic cleavage. Dig into some mineral field guides, or jump on the web and search. Get yourself a hand lens, 10X works well. Crush some of your ore, get it in a small gold pan, in natural light, and examine the ore with the lens. Gold particles will have a soft yellow luster, pyrite will be bright and shiny with cubic, square corners.

The best case scenario is oxide. These ore types are the easiest and cheapest to mill. Oxides are those brightly colored rocks or rock formations you see. Some are pastel colors, like pink, mauve or orange. Typically, the reddish and orange colors indicate the presence of iron oxide, and the darker the red, usually the higher the iron content of the mineral. The pinks and purples are usually a mineral with some manganese, and the tan to brown you see will most likely be limonite, another iron oxide. Bright Kelly green is most likely a nickel oxide, a good indicator for platinum, and silver bromide is usually a pea soup green.

Why do you care if the rocks are pretty colors or not? Oxide ore deposits are considered "free milling". Unwanted elements, such as iron, may be present, but they are not alloyed with the values you are after. The gold will be yellow metal, not an alloy with iron pyrite or other elements. The values will be much easier to recover.

Some good books to help you start understanding geology and rock types is the "Roadside Geology" series, hopefully there is one available for your state. If you want to get into the geology of a specific mine or mining district, get on the web and locate the Bureau of Mines website for your state, or the state where your mine is located. There are dozens if not hundreds of mineral identification books out there.

As a matter of course, the Bureau of Mines compiled reports of mines that historically operated, and the information is usually available at no cost. A web search can also find various books that describe all of the mines in a mining district, which should be of great interest to you since they will describe the mineralization of the district. The same applies to placer mining in the same districts.

The desert southwest is quite often overlooked, which is a mistake. The vegetation is sparse. The "desert varnish" on rocks is usually iron or an iron manganese stain on the top surface of the rock. To see what the rock is, turn it over. Or break it with your rock pick. A drop or two of any mineral acid will also dissolve the varnish. The other advantage to the southwest is that surface ores are oxidized. The high temperatures in say, Arizona, have been gradually roasting the rocks (and people) for many millennia. If you are mining in the desert southwest, you will encounter a transition zone where the ore transitions to a sulfide, gradually at first, then heavier. The deeper you mine, the heavier the sulfide. In northern areas, such as the Colorado mineral belt, it is common to see sulfides right at the surface.

Oxides are also the easiest to fire assay, and at this point, the most important information you can have is the value of the ore. A simple fire assay is fairly quick, relatively inexpensive, and normally for gold and silver.

"Ah", you say, "I had a piece of this ore assayed, and it assayed 100 ounces per ton, so I'm going to use my Cousin Fred's cyanide plant to process the ore". Already you are in trouble for a whole bunch of reasons. First off, one piece of rock isn't a representative sample of your ore, and secondly, if your ore is actually 100 OPT (ounces per ton) of gold, cyanide is the wrong way to go. Way, way wrong on many levels.

The point is, at this stage, you should be gathering information to ascertain the correct milling process that will give you an acceptable recovery for your ore. And no, you never get it all. In fact, if you can recover 90% or more of the available gold and silver, you have done very well.

Your search for information should begin with the basic fire assay from several labs. This provides an estimate of the value of the mineral you have. Is it worth processing? For instance, if you have 5 OPT of silver, you will never even come close to a profit on your ore. On the other hand, if your 5 OPT are gold, well, a different story. You now have a chance of making a decent profit.

There are some basics you have to understand at the beginning. First, mining is a business like any other. The idea is to make a profit. Second, what is your cash cost per ounce? What will it cost for each ounce you produce, including all expenses? Obviously, if your cash cost exceeds the value of what you produce, you are losing money, and do not have a viable process. If gold is at $1500 per ounce, and you are spending $1600 per ounce to produce that ounce or those ounces, some big changes are required, or you will be bankrupt. Keep in mind that your cash costs per ounce can be quite high for complex, refractory ores.

Another very basic common concept in mining is this: *Heads minus tails equals recovery*. Heads being the assay of the head ore, in ounces per ton, as it will enter the milling circuit. Tails being the assay, in ounces per ton, of the waste, or tailings from the milling circuit. If your head ore assay is 10 OPT gold, and your tail assay is 5 OPT gold, that would be a 50% recovery. This would tell you that you have serious problems in your mill process. The point is that a tail assay can be even more important than a head assay. Get the numbers you need to make necessary operational decisions.

When you are a mill operator, you will quickly learn to pay attention to the numbers. A mine manager and mill superintendent will be looking at the numbers several times a day, the sooner the better. The numbers as described above justify the existence of the mill and the mine. If you work in the mill or mine, well, maybe you should ask about the head and tail numbers, since it may be your job at stake.

Another concept you should understand is that the metal you sell is not pure gold. You will produce what is called Dore'. This is a combination of all the precious metals that have been recovered from your ore, as well as some base metals. Copper tends to follow the precious metals as well. Platinum group metals will probably be in the Dore' as well. If your mill is really sloppy, the base metals can actually rise to the "penalty" level, and the refiner will charge an extra percentage for refining your metal, or Dore' bars. There will be more on this later in this book.

Contaminants

Now you need to find out what else is in the ore. There is an easy way to do this. Send a sample of your original representative sample out for a spectrographic analysis. Get the big one, 58 elements. That will give you the most bang for your buck. Does your ore have cadmium, beryllium, arsenic, mercury, gallium, uranium, or thallium in it? How about lead? Where are you going to mill your ore? Each state has limits on certain elements that may be processed, or liberated from a natural deposit. Some toxic elements will not be permitted at all.

Mercury is now a legal dirty word. There are federal anti-export laws on mercury in every state. Do you have mercury? How will you store it forever and ever? Thallium was a famous rat poison (four legged) for many years. Do you really want to mine this? Even naturally occurring lead can be a permitting issue. What will you be exposing your employees to? How will these contaminants affect your mill circuit?

Another reason to spectrograph is to understand that some of these elements can contaminate your dore' and become a penalty element. The refiner has to deal with whatever elements that follow the precious metals. Something toxic, such as cadmium, will create a disposal problem for the refiner, so the costs are passed down to the producer of the metal. The allowances on penalty elements are different from refiner to refiner, so it may be worthwhile to shop around. One penalty element most people don't think about is copper. And copper is common.

Get the spectrograph before you spend a lot of money or time on permitting a mill site. Don't be naive. Any agency that would permit a milling operation will certainly require a spectrographic analysis prior to permitting. Better to know ahead of time what you may have to deal with in the permitting process.

The good news is that you may have a rare earth or something of value you weren't even looking for. It is also a good idea to assay for the platinum group elements to eliminate or verify them at the beginning. The platinum group elements (PGE's or PGM's) can be a penalty element as well. Normally, the marketable ones are platinum, palladium, and rhodium. Osmium, iridium and ruthenium are not considered commercially significant. Normally, they will be the by product of processing another element. Think about gypsum, calcium sulfate. There's an unending market for such industrial minerals. You should identify every major mineral deposit on your property. There are a lot of geologists out there that will gladly assist you in the process. Did you know that uranium ores are 40 times more common than silver, and 500 times more common than gold?

You have the spectrographic analysis in hand. A few nasty things in the report, but all at trace levels. The report is by element, in parts per million (PPM) or parts per billion (PPB). Remember that 1,000 PPB are one (1) PPM. And note that 1% equals 10, 000 PPM. If the element in question is a precious metal, convert PPM to OPT by multiplying PPM by .029166. If you see a "<" sign, that is "less than", ">" is "greater than".

Element Name:	Symbol:	Element Reported in:	Notes:
Aluminum	Al	%	Interferes with chemistry, will sell
Antimony	Sb	ppm	Toxic, contaminate, marketable.
Arsenic	As	ppm	Toxic, some industrial uses.
Barium	Ba	ppm	Alkaline earth metal
Beryllium	Be	ppm	Rare metal, toxic
Bismuth	Bi	ppm	Rare metal
Boron	B	ppm	Rare metal, toxic
Cadmium	Cd	ppm	Rare metal
Cerium	Ce	ppm	Rare earth
Cesium	Cs	ppm	Rare metal
Chromium	Cr	ppm	Rare metal
Cobalt	Co	ppm	Rare metal
Copper	Cu	ppm	Marketable
Dysprosium	Dy	ppm	Rare earth.
Erbium	Er	ppm	Rare earth.
Europium	Eu	ppm	Rare earth.
Gadolinium	Gd	ppm	Rare earth.
Gallium	Ga	ppm	Rare metal, can be toxic as Arsenide
Germanium	Ge	ppm	Marketable. Rare metal
Gold	Au	ppm	Marketable.
Hafnium	Hf	ppm	Marketable, Rare metal
Holmium	Ho	ppm	Rare earth.
Indium	In	ppm	Rare metal
Iridium	Ir	ppm	Marketable, alloys with Osmium.
Iron	Fe	%	Interferes with chemistry.
Lanthanum	La	ppm	Light rare earth element.
Lead	Pb	ppm	Toxic, marketable.
Lithium	Li	ppm	Rare metal, very marketable.
Lutetium	Lu	ppm	Rare earth.
Magnesium	Mg	%	Marketable, alloys with Aluminum.
Manganese	Mn	ppm	Slags silver in nature. Rare metal
Mercury	Hg	ppm	Toxic, some industrial uses.
Molybdenum	Mo	ppm	Marketable, Rare metal
Neodymium	Nd	ppm	Rare earth
Osmium	Os	ppm	Toxic, marketable, Rare metal
Palladium	Pd	ppm	Marketable. Rare metal
Phosphorus	P	%	Marketable, shortage coming.
Platinum	Pt	ppm	Marketable. Rare metal
Praseodymium	Pr	ppm	Rare earth.
Rhenium	Re	ppm	Marketable, many industrial uses.
Rhodium	Rh	ppm	Marketable. Rare metal
Rubidium	Rb	ppm	Marketable, Rare metal
Ruthenium	Ru	ppm	Marketable, hardener for Pt, Pd.
Samarium	Sm	ppm	Rare earth.
Scandium	Sc	ppm	Rare earth.
Selenium	Se	ppm	Rare metal
Silicon	Si	%	Rare metal
Silver	Ag	ppm	Marketable

Sulfur	S	%	Marketable
Tantalum	Ta	ppm	Rare metal
Tellurium	Te	ppm	Rare metal
Terbium	Tb	ppm	Rare earth.
Thallium	Tl	ppm	Toxic, Rare metal
Thulium	Tm	ppm	Rare earth.
Sodium	Na	%	Marketable.
Strontium	Sr	ppm	Alkaline earth metal
Tin	Sn	ppm	Marketable.
Titanium	Ti	ppm	Rare metal
Tungsten	W	ppm	Rare metal.
Uranium	U	ppm	Rare metal
Vanadium	V	ppm	Rare metal
Ytterbium	Yb	ppm	Rare earth.
Yttrium	Y	ppm	Rare earth.
Zinc	Zn	ppm	Interferes with chemistry.
Zirconium	Zr	ppm	Rare metal

Take a look at a sample report above. This layout is to make the report easier to understand. Normally, the elements won't be sorted alphabetically, and the note column will hopefully, create a little curiosity about the element. Access the World Wide Web, and check out the element in question. Google it! That works every time.

Note that some elements are labeled toxic. You should pay close attention to these elements and the concentration present. How you intend to mill ores containing such elements will be part of any permitting process, even if you don't recover them.

Rare Metals

While looking at your spectroscopic analysis, take a look and see if you have any rare metals. While all that glitters isn't gold, there are other valuable elements out there.

There are, no doubt, some physicists and chemists that will take issue with the Spectrographic report above. Before you do, read "Rare Metals Handbook" by Clifford A. Hampel, First and Second editions. The point is that if you have a spectrograph done on your ore, and the numbers come in high on the rare metals, or rare earths, you need to do your homework. Buy the book, read the information. *The book even tells how the element is recovered, and shows circuits.* It doesn't get any better than that.

Even the old standard, Uranium is still active. Mining conglomerates have been quietly buying up old Uranium properties the last ten years or so. Why do you suppose they would do that?

To find the Rare Metals Handbook, go to the Miners, Inc. website. It is www.minerox.com. Or call 800-824-7452. They have a wonderful selection of books, and also produce one of the best catalogs out there, with a different mineral specimen on every different catalog cover. They are a full line supplier of mining accessories, so you will enjoy their catalog.

S. E. M Microprobe

Another method of analysis that can be quite useful at this point is the Scanning Electron Micro probe. Or "micro probe", or "SEM Micro probe". Think of an Electron Microscope that can scan for all elements. The device is a long way from a microscope, and very complex.

Normally, a cut sample of ore around 1.5" X 2" is required for analysis. And yes, a tile saw can be used to cut the sample, or if you have a lapidary shop in your area, they can cut the rock for you. The Micro probe can also scan assay beads, or Dore' from a smelt. A small piece of pin tube sample can be scanned as well. Contact the operator or institution for requirements. Micro probe will also inform you of the penalty elements in your Dore'.

Bart Cannon Micro probe (www.cannonmicroprobe.com) is still out there, and Arizona State University has Micro probe services available. Check out http://le-csss.asu.edu/node/241 Also inquire about the turn around time on samples. Two weeks is not unusual. Contact information is on the web page.

Micro probe information is important because a lot of people, especially beginners, will see a dense grey metal on a concentrating table, or in their pan. Some self-professed idiot will suddenly say "Oh my God, you've got platinum" (or PGM's or PGE's,) whatever their buzzword is, and everything goes to hell quick. Ever seen specular hematite? Nickel? Iron? There are about a hundred different elements that will be grey, or gray. Don't overreact, analyze the metal, get the answer.

The chances there is elemental, metallic platinum in your sample are slim to none, and Slim's out of town. There is metallic platinum being mined on this continent, but the occurrence is actually rare.

There is one last analysis you should have at this point. This is a Leco analysis. The Leco machine actually used high temperatures to oxidize any sulfide present, and capture the percentage of sulfur in the vapor stream as well as carbon. These two numbers (sulfur and carbon) will become very important in any process to recover your gold if the circuit is anything beyond a simple gravimetric circuit. Have the Leco numbers done if the sample has any hint of sulfide or carbonate.

Keep in mind, sulfides are not necessarily the "kiss of death" for your mine. Some sulfides, such as Marcasite, do not alloy with gold, but are a positive indicator for gold. Isolate and assay the sulfides separately and see where the values are. Most sulfides normally carry some gold.

Chapter Eight

Gravimetric Devices

These are the most simple and easy to use circuits, and are useful for heavy, dense elements. The idea is to get particles of a similar size in water, and let gravity do the work. At this point, the old "KISS" method (Keep It Simple, Stupid) should be foremost in your mind. Understand that you have to have a way of grinding your ore, and a lot of the equipment out there will be highly touted by the manufacturer. The proper grind is the one thing you have to have for your mill to be successful. In real estate, they say "location, location, location. In mining, we say "classify, classify, classify". For a reason! An 1/8th inch particle of rock will displace a +100 mesh gold particle every time. This will be in any gravimetric device. Gravimetric methods are all about particle density. You don't have to be a physics major to understand the concept, so make sure you classify your ground material.

Concentrating Tables

Concentrating tables or just "tables" have been around a 100 years or so. Simple, easy to set up, and easy to use. Many different brands, such as Deister, Wilfley, and a more recent favorite, the Tra-Lite table. The Tra-Lite is no longer being manufactured, and the used ones are in demand. Deister and Wilfley are still being made. The Tra-Lite tables have a fiberglass deck, and have a considerably faster drive than other brands. Wilfley and Deister both have replaceable covers for the decks in different configurations. If you are contemplating the use or purchase of a table, Google the manufacturer. There are pictures of deck configurations, videos, and setup instructions.

Some are "bump" tables, with an eccentric of some sort bumping the deck of the table, providing a lateral motion of about ¾ of an inch or so. The table will have a series of ridges, or grooves to capture the heavier minerals running the length of the table.

The most common deck configurations have three grooves running the length of the table. There is one last groove at the very bottom of the deck, usually a safety groove in the event of a feed surge. The three grooves are referred to as "cuts", with the top cut being the "high grade cut" then "seconds" and "middle cut" or middlings. What material passes all the way across and off the table is "tailings", or waste.

The header box at the top side of the table is where either a slurry or dry, pulverized, *classified* ore is fed to the table. A slurry is far better than dry ore since it has already been wetted, and the gold won't float away before it can be wetted. Use a surfactant, such as Jet Dry, a non-sudsing dish rinse, or Lysol or even Formula 409 in small quantities. If the surfactant makes suds, your gold will float. If you are feeding the table from a slurry tank, start at 30% solids, and keep the tank agitated and add your surfactant in small quantities to the tank.

In a good free milling ore, it is not unusual to see a solid line of gold come down the top cut, and off the table, where it is captured in a container. In the case of gold, if the table is operating and set up properly, this cut can be dried, fluxed and smelted to Dore' as is. The second and third cuts are usually captured and put across the table separately, or across a smaller table as a "clean up".

If possible, run your table in natural sunlight, or the equivalent. Artificial light makes metallic gold particles hard to see, especially fluorescent light.

Wilfley, Deister and others have a positive mechanical drive with an adjustable stroke, and do not bounce. There is less wear on the positive drive mechanisms than the bounce tables. Bounce tables are usually a lower end device due to the lower manufacturing costs of the drive mechanism. Increasing the stroke on a bounce table can be done in some cases by adding weight to the eccentric, however this tends to shorten the life of the drive mechanism considerably.

Most problems with tables come from improper set up. Start with a *level* concrete pad. Secure the table to the concrete pad, and level the top. Make sure you have adequate, *clean* water. A 4' X 8' table will use one inch of water at the header box. If you are feeding a slurry to the table, less water may be required. Drop the lower edge one inch, and start the table and water to the header box. The desired effect is a "fan" almost across the table. If waste is traveling the top cut, or the fan crosses the table, drop another inch. Watch for dry spots as the table runs, and adjust the water to the table accordingly.

Once the table is operating correctly, leave it alone. If some expert wants to adjust it for you, make sure he's an expert. Once the table is operating correctly, you will be able to run a fairly broad range of ores with the same settings.

Never ever set anything on the table, it will damage the top. Likewise, keep your hands off the table, the oils from your hands will cause surface tension and the gold will float. If you have to touch the top, wear vinyl gloves like doctors do.

There are a lot of cheap imitations out there, you should remember that you get what you pay for. There are tables with magnets, and all sorts of gizmos. Make sure you test any table with a 100 lbs of classified ore.

Remember that tables have a huge number of uses, on of which was around for many years. The separation of tungsten ores, as well as barite was done by table for most of a century. Tables are also used extensively in the coal industry.

Keep in mind a replaceable top is a good idea, ore at the very least, a urethane or fiberglass top. Make sure the top won't decompose in strong sunlight. The idea would be to use a table capable of a fine cut, and avoid amalgamation at any cost.

Gemini Tables

These tables are in fact, refiner's tables. They are eight feet long, and are "coffin" shaped. The top is urethane, with very fine cuts. They are center feed, and the water is center feed as well. The cuts run the length of the table. The slurry is washed down both sides of the table making one think the table is high capacity, and it is not. The newer tables come in three sizes, and are capable of 60 lbs per hour to around 900 lbs per hour. There are three high grade cuts on this table. The table has an adjustable mechanical drive with a fairly short, fast stroke.

The purpose of this table is to separate materials that are close in specific gravity. They are great for cleaning up concentrates from any gravimetric system. They are also excellent for beach sands. The table is easily capable of separating black sands from any concentrate. You won't believe how fine the gold is that they can recover.

The key to operating these tables is patience. Always install them on a level concrete pad, and bolt the table down solid. Make sure there is adequate *clean* water, and the feed material is classified. Adjust the water flow down the center of the table to eliminate any dry spots. Never touch the top of the table with your bare hands, and never set anything on top of the table. If the table is set up outdoors to take advantage of natural light, cover the top when you are done for the day. Never overfeed the table, and the use of a table feeder is strongly recommended.

If the table is properly set up, the cuts will produce smelter grade material. It is simply a matter of drying the high grade concentrate, fluxing it in the crucible, and firing to Dore'. This is why the table is referred to as a refiner's table, or a "clean up" table. Other than a surfactant, no chemicals are required with this table.

There is a four foot "lab" table available, and they will work very well for cleanup of bowl concentrate, or other classified material.

Contrary to popular belief, these tables are still being manufactured in Australia, and sold in the US. Go to www.mineraltechnologies.com and look at the brochure. The salesman to talk to is Peter Barker in Florida. Call 904-827-1696. There have been significant improvement to the top material, drives, capacities, and top configurations.

Micron Mill Wave Table

These tables are manufactured and sold by Action Mining Services in Sandy, Oregon as well as their distributors around the world. These tables work on a completely different principle than most tables, and are now available in a variety of sizes. They are easily capable of generating a smelter grade product in the correct configuration. The Micron Mill Wave Tables do not require any chemicals other than a surfactant.

The best way to get the information on these tables is to go to Action Mining's website, www.actionmining.com, and look at the table catalog. All the pertinent information is there. You can also call 503-826-9330. There are also You Tube Videos of the tables working.

Contact Action Mining for information, and see if they won't run a test on your ore. As with any test, you should retain your tails for analysis if possible.

Action Mining have always been innovative in the mining business. Download the catalog while you're on the website, you might be surprised at what they have available.

Jigs

Jigs are another gravimetric concentrating device. Jigs are typically a square box with a flexible diaphragm on the bottom. The diaphragm is connected to a crankshaft that travels vertically up and down, and the stroke is adjustable. The top has a screen, and it is covered by a layer of shot. Typically, the shot is steel, and is about the size of a BB. The slurry of ore is fed on the screen, and with the box full of water, the diaphragm is moved up and down, creating a surge that expands, or flexes the shot bed. The heavy particles pass through the shot bed, and collect in the tank above the diaphragm. Theoretically, the lighter and oversize portion of the slurry wash across the shot bed, and are waste.

There are some serious problems with jigs. The first being the amount of water required to operate the jig. A hole in the screen can be a disaster, as well as a ruptured diaphragm. Some serious test work is required to determine the correct size of the shot. The setup and tuning of the machine is tedious at best. There are many variables that must be set just right for an efficient recovery. Any variation in water flow can decrease recovery.

Some operators swear by jigs, and there are many out there. There aren't many in use in the desert environment due to the high water requirements. They have been used successfully to concentrate placer gold, tungsten ore, and even barite.

Concentrating Bowls and Centrifuges

These are gravimetric devices, operating on the same principle as a table. The difference in specific gravity of the precious metals as opposed to the specific gravity of the waste is the key. The added twist is the turning motion of the device, which increases the effects of gravity. The weight of the particles is increased relative to the speed of the rotation. Water is used to carry away the lighter waste particles in the case of the bowls, and most other devices have a spiral, or other mechanical device to help evacuate the waste from the machine.

The same rules apply for these devices as any other gravimetric device. Classification, again, is the key. The closer the particle sizes are, the more efficient the device and recoveries will be. A surfactant is also a useful addition to the slurry fed to these devices.

Neffco and Knudsen bowls are pretty common, and have been popular for years. Neffco is out of business, so take a look at Sepro "Falcon" Centrifuges. Go to www.seprosystems.com The Falcon Centrifuges are the latest generation of hydrofuges, with speeds up to 600 RPM.

The big advantage is the volume that they will handle, as compared to a table. They are normally used as scalpers, capturing all the heavier elements. The concentrate from bowls is then tabled to clean it up. Neffco bowls have a J riffle on the inner side of the bowl that works well, but Neffco bowls are getting hard to find. Knudsen bowls have adjustable blades in the bowl, and can be tricky to set up. Typically, the bowls spin at 100 RPM or so. Normally, bowls are operated in pairs, with the second, lower bowl being a "surge" bowl in case the values are accidently flushed out of the primary bowl. Constant, steady feed is important.

Centrifuges are out there in about every configuration, and about every speed one can imagine. Some are very slow, and are basically a nearly horizontal pipe with a spiral molded inside of it. The recovery is adjusted by raising or lowering the end of the pipe. These are basically the equivalent of a concentration table with a riffle that spins, and collects the gold. Beware the super 10,000 RPM machine, the maintenance costs are going to be horrifying.

An offshoot of all this is the spiral, which is sometimes referred to as an "automatic" panner. These devices tend to be rather sloppy, and the material fed must be classified for a decent recovery.

Manufacturer's will claim recovery of "micron" gold, or some such. Best to test the machine on your own ore, and verify the manufacturer's claims. If you really do have micron gold, you had better look at one of the chemical processes to get a decent recovery. If you suspect extremely fine gold, pan the coarse gold out of your sample, and have your assayer do a hot cyanide shake on the pan tail. See what you missed, if any.

Free gold in a sliced ore sample.

Chapter Nine

Miscellaneous Devices

Placer Plants
There are many different brands of placer plants out there. You can do a Google search, or go online to the International California Mining Journal, www.icmj.com, and peruse the ads. This will give you an idea of the prices and capacities that are available. Once your pilot work is out of the way, you can decide what you can afford, and the capacity of the equipment you can afford. Keep in mind that the larger the plant, the more water you will need.

Some refer to these devices as "wash plants", but the actual process goes far beyond washing.

There are small, shovel fed systems clear up to 100 or more tons per hour. Before you decide anything, establish your water source. Decide or find out if you will have to recycle water. Extra acreage will be required for tanks if you recycle, and your plant will have to be stationary. Logistically, constantly moving big tanks can be a nightmare.

Will you be able to transport the gravel to the plant over a long period of time? Or will you have to move the plant? Do yourself a favor, go to the manufacturer and take a long look at the equipment. Also, any reputable manufacturer will tell you where to see their product in operation. This will also give you access to the operator, who will be able to answer your questions.

There are used systems out there. Will the manufacturer honor their warranty on used equipment? How much to bring the system into working order? Check carefully for dozer dents and loader dings. If there is damage to the trommel, don't buy the machine. The trommel has to run true, so it would have to be replaced. Find another one.

The same rules apply for all equipment. Do your homework.

Table Feeders

It seems that overfeeding is a continuing problem with most gravimetric devices. If you are using a table, bowls, or whatever, find the appropriate feed mechanism, and use it. There are a variety of table feeders our there, some a simple hopper and conveyor to an auger mechanism, and then there are vibratory feeders. One simply loads the feed device at regular intervals and the feeder maintains a constant feed rate. No surge.

Bowls are sensitive to surges, and a smart operator runs them in pairs for this reason. Any values pushed out of the first bowl are captured in the second bowl, and saves a major recovery effort from the tailings. A smart operator will also have a table at the end of the circuit for a final recovery. After the bowls have been cleaned, the table can be used to clean up the bowl concentrate, and if the right table is being used, such as the 4' Gemini, smelter grade material is recovered. An assay of the tails off the table will tell you how well your circuit is working. Build a composite sample, and check your circuit regularly.

Just remember that if the power fails, then kicks back on, you could lose some serious values, so plan for surges, feed failures, and so forth.

Magnetic Separators

The nemesis of miners worldwide is iron. Logical thinking would dictate that iron, being magnetic, could be removed with a magnetic separation. Maybe, maybe not.

A magnetic separator can be invaluable *on some ores*. There are many types of magnetic separators out there. Belt and drum separators are the most common, and both can be found that will run wet or dry. If you have a concentrate, and you are trying to get rid of the black sand, first pan out some of the black sand or do a magnetic separation by hand, and have it assayed. If you have a coarse black sand, there is a good possibility it will carry gold. Always assay any portion of any separation, magnetic or otherwise to ensure you don't discard any values. Sometimes, a simple re-grind to a finer particle size will do the trick.

Usually, a carefully adjusted concentrating table will do the trick Test your ore on a table before making a large investment, sometimes the simple things work best.

With the advent of the rare earth "super magnets", the prices of this equipment is considerably more expensive than the older, permanent magnet type. As always, have the manufacturer test a 100 lbs or so of your ore prior to purchase. Assay both portions to make sure a magnetic separation will actually upgrade either the magnetic or non-magnetic splits. Don't be surprised if the separation process shows no gain.

One place magnetic separators are under utilized is on beach sands. Beach sands can carry respectable amounts of gold, and some will have elemental platinum group metals present. Your author has spoken with many prospectors that were amalgamating beach sands. It never occurred to these folks to try a magnetic separation. *Never* amalgamate. The mercury contaminates everything it comes in contact with, and there are no legal methods of disposal. Getting caught amalgamating is guaranteed financial ruin, and these days, jail time is a very real possibility.

Action Mining has a hand separator, and other different types of magnetic separation devices. Go to www.actionmining.com, and check the catalog.

Yellow Iron

When you bring heavy equipment to your project, keep the inexperienced off the machinery. Many a decent mine or prospect has disappeared after someone playing on a bulldozer buried it. Water is also an issue. Water can be inadvertently diverted on or off your property, and since the whole country is pretty much wetlands according to the EPA, do you want to go there?

Also at risk will be your sample markers, claim stakes, innocent bystanders and everything else you can thing of. A D-10 won't even slow down going over a passenger car. Hire professional operators, and communicate well with them. Work off maps, and remember, this equipment has GPS, and laser alignment devices for accuracy. A professional operator will use these features. Your brother in law won't have a clue.

Chapter Ten

Crushing Your Ore

Jaw and Cone Crushers

How about this? If you have 50,000 tons of ore, that doesn't mean you need a 10,000 ton a day mill. How about keeping your expectations realistic, build a pilot mill, and process 100 or 500 tons a day until you can afford that bigger mill? As with any other business, cash flow is the name of the game. That is the purpose of this book. To educate you, and provide enough knowledge for you to succeed.

The first step is to figure out the best, most economical method of making little ones out of big ones. Normally, the primary crusher is a jaw crusher. This device has a stationary plate, and a movable plate. The toggle plate allows the movable plate to lift, then move forward as the plate descends, and applies enormous pressure on the rock, which breaks it. Size is controlled by adjusting the opening at the bottom of the crusher. Go to You Tube. There are videos and plans for all types and every size of jaw crusher.

Normally, a jaw crusher is described by the width of the jaw. This can be anywhere from lab sized equipment (4" X 6") right on up to and over 36"X 30"for gravel pits, and large mining operations. Maintenance is straight forward, and from time to time the jaws are hard faced (welded) to replace the worn away metal. Daily lubrication is a must. They are by their nature and design, very heavy and require a fairly large electric motor. Do you have adequate power at your location, or mill site? Usually, the minimum is 220 volts 500 Amps or more. Standard is typically 440-480 volts, and several thousand amps. All crushing equipment is rated in tons per day, or better yet, tons per hour. The manufacturer will claim tonnage based on "bull" (barren) quartz. That's the pure white rock crushed up for landscaping applications. That's a hardness of seven on Mohs scale. A diamond would be 10. All jaw crushers have large, heavy flywheels. Once the flywheel is up to speed, they maintain speed, and insulate the electric motor from the resistance of the jaws as rocks of assorted hardness are crushed.

Most jams are caused by over feeding the crusher, and can easily be cleared by powering down the motor, and turning the flywheel the opposite direction. This pushes the jammed rock up, and allows the rock to be removed. Remember electrical lockout and tag out. The last place to be is anywhere near the flywheel, or have a bar in the jaw when the machine starts. If you do, well, be prepared to be an industrial statistic. Also, slow down the feed! A feed hopper is a really good idea, as this maintains a steady feed to the crusher, with out overfeeding the jaw. Just dumping loader buckets in a jaw crusher is not a good idea. Think about using another feed hopper before the secondary crusher, as well. The primary crusher, being larger, will have a higher capacity than the secondary crusher.

If your jaw crusher is crushing from a range of 24" rock to 4" rock, your next step is a secondary crusher. This can be another, somewhat smaller jaw crusher or a gyratory, or cone crusher. Cone crushers work by offset wobbling a heavy metal skirt around a heavy case. The skirt is tapered, and the size is controlled by raising or lowering the skirt. The driven portion of the crusher is in an enclosed housing, and usually oil filled. This crusher can take 4" rock down to 1" in one pass. Again, go to You Tube, and watch a few videos. The cone crusher is considered high maintenance, and jams can be a nightmare to clear. The same safety rules apply as with the jaw crusher.

Usually, the arguments pro and con are pretty much 50-50 for and against each crusher. The jaw is easier to maintain, and easier to clear if a jam occurs. Never look inside any crusher when it is operating, even the small ones. Some really hard rocks will pinch, and accelerate back out the top, removing your face, hand or whatever is in the way in the process.

Another point to consider is that you will need a feed hopper and conveyor to feed each crusher. The trick is to provide a constant, steady flow of ore to the crusher. Feed the primary jaw at the capacity of the secondary crusher. This saves a lot of jams. You can also feed the crushed ore from the primary crusher to a screen deck and screen off the undersize to lighten the load to the secondary crusher. You'll need another feed hopper as well as the screen deck.

Screen decks come in all sizes, and the screens are available in all sizes. They can save a lot of time, and are all vibratory with a pan below the bottom screen to capture the screen fraction you need to separate from the material going to the secondary crusher. This material is conveyed past the secondary crusher to the next step in the process.

From the secondary crusher, the next step (tertiary) can be a set of rolls, if the size of the ore needs to be reduced further, or to a ball or rod mill, or a SAG mill. A set of rolls can be quite handy with most ores. Rolls are simply two solid metal cylinders running face to face horizontally. They have axles that have heavy, adjustable springs mounted on the shaft. The cylinders are usually driven be an electric motor, and the faces of the cylinders are hard faced to offset wear. Some rolls are built to the same inside diameter of steel pipe. When the pipe is worn off, the old pipe is removed, and another section is slid on the roll. This method is considered the easiest for maintaining the rolls, which will wear rapidly in harder ores.

You have now made it to the fine grinding stage, but there is another way to do all this, and why small to medium sized mines don't do this is a mystery. Why not save some money? Why work harder? Try a gravel plant.

The Gravel Plant

Buy, rent or lease a gravel machine. The beauty of these machines is that they do it all. They typically come with their own power plant, and it is usually diesel. They are designed to be portable, and are pulled to location by semi tractor. They have single or multiple crushers, screen decks, and conveyors on board. Everything you need in one mobile unit. They may or may not be affordable for your project. Keep in mind that this equipment can be leased. Compare the cost of building a crushing circuit, then look at purchasing or leasing a gravel plant.

If you have a 50,000 ton pile of ore, check the rating on the gravel plant. If the manufacturer rates the plant at 1,000 tons per day, the next question is what type of material, and what is a day? If the 1,000 tons per day is in bull quartz, and a day is defined as eight hours, that's a lot of ore that can be processed per day. If a day is 24 hours, well, that's roughly 41 to 42 tons per hour, not near as impressive.

Is it worth buying used? If you are really mechanical, or can hire a good mechanic, well, maybe. It can be very expensive to rebuild this type of equipment. If you lease, look at getting an operator before you start. How about just hiring a contractor to come in, process the pile, and leave? How much per ton would this scenario cost? Make sure that whichever way you plan to go, you permit accordingly. No municipality will permit a gravel plant within city limits. Noise and dust are going to stop you from a permit wherever you go in a municipality.

Realize you will have a somewhat smaller pile of ore, and typically a gravel mill will produce 1/8th minus material. You should know where the values are at this point, so if the values are in the -20 mesh material, you can concentrate the ore pile by simply screening out the oversize with a screen deck. You would screen off any material from -1/8th inch to +20. If this were to reduce the volume of your or pile by 10 or 20 percent, you would still have the original values in a smaller volume of ore. This would be an inexpensive method of concentrating your values. Verify this concept with a simple re-assay of the ore pile, don't ever assume anything.

If this trick works, your 50,000 tons will now be 40,000 or 45,000 tons of ore, with an increase of .500 to 1.000 OPT of gold in your ore pile. You didn't create any gold, you simply concentrated the values to a smaller volume.

A typical Haul Pak. This type of vehicle may be required by MSHA for safety reasons.

Chapter Eleven

Grinding to Mill Requirements

Impact Mills

Once you have the crushing done, you will need to grind to whatever mesh your lab work showed to be the point of mineral liberation.

The are several machines out there that are a combination jaw crusher and impact, or "hammer"mill. The impact mill will consist of a rotor with paddles on the rotor, and typically, impact plates that the rotor sends the ore into at high speed, breaking the ore into smaller fractions. The rotor typically is horizontal, and the shaft is usually three inches or more in diameter. Impact mills are fairly inexpensive to build, and deserve some consideration. Some of the newer mills are configured with a vertical shaft, yet basically function the same a the horizontal shaft mills.

The main problem is that the impact mills currently on the market *do not classify*. Yes, they break up the softer portion of the ore, and generally grind finer than necessary. Sometimes several passes through the machine may be required if the values are in the harder portion of the ore. *Your author has never seen an impact mill that grinds to a given mesh size in one pass.* Ever. The manufacturer may claim the mill does grind in one pass, however a quick test with a sieve of the size you desire will prove or disprove the manufacturer's claim.

Some manufacturer's claim 500 mesh in one pass. Brilliant! You will have created slimes, and pretty much made all your gold into flour gold. This is a major problem whether you intend to use chemistry or gravimetric methods. You can't see gold any finer than 375 mesh. Why would you grind this fine? In fact, anything finer than 400 mesh in mining is considered slimes. A -400 mesh particle is 0.0015 of an inch, or 37 microns. A classic example of over grinding. The particles will suspend in any solution, and will require treatment to collect them. Don't go there.

If you know your ore has to pass 80 mesh (-80) for mineral liberation, take one 80 mesh sieve and some ore to the manufacturer. After milling, take a scraped level cup of the ore, and put it through the sieve. Catch the undersize. What remains on top of the sieve is oversized, and then you will know. Many people buy these "all in one machines" and don't test them, only to learn they do not do what they need. These machines are usually a small jaw feeding an impact mill.

Either way, you will have to separate the screen fractions to your mill specifications as defined by the screen fraction analysis previously explained. So you will be buying more equipment.

A screen deck or similar device will be necessary to classify the ground ore. Cyclones can be used for this purpose, and there are models out there that will work wet or dry.

Typical output for an impact mill is minus ⅛ inch. Try screening out some of this oversize, and have it assayed. If there are low or no values, just screen it off, and proceed to the next step. Never over grind your ore. If there are economic values in the oversize, you have to run it again, and maybe again, depending on the hardness of the ore. Remember that your assay report has told you where the values are. Grind until the values are liberated.

It's pretty obvious that dust is an issue. If you have to run wet, make sure you have access to adequate water. If you are using an impact type mill, note that running wet will require more horsepower.

Dry drilling has been illegal for many years, and wet drilling was introduced to prevent silicosis. The same issues apply to dry grinding. The EPA has become more and more strict on fugitive dust. One complaint, and you will be under a microscope by the EPA, and that isn't a fun place to be. Many very small municipalities have been forced to pave, or cover dirt streets for this very reason.

Vertical Shaft Impact Mills

Normally, this type of impact mill is used in gravel operations. They could be quite useful in a mining operation with certain types of ores. A variation of this mill uses rock as the anvil, saving wear. This type of mill can be purchased with adjustable anvils, with produce the particle size required in a grinding operation. These mills also come in various portable configurations. Be sure to test your ore before purchase. These mills are popular in gravel pits that produce highway aggregate. If you plan on grinding dry, be prepared to deal with the dust produced by the mill.

Autogenous Mills

An autogenous mill is basically the same as the SAG mill described below. The difference being that an autogenous mill does not use balls. You could say the rocks are grinding the rocks. Otherwise, the same process applies. Autogenous mills can have a huge advantage with the right ore types. Autogenous mills can be vertical or horizontal, however horizontal seems to be the most popular.

SAG Mills

Once you have passed the primary and (if needed) secondary crusher, it's time for some fine grinding. One of the better ways to start the process is with a SAG mill. SAG stands for "semi-autogenous grinding". What that means is that the ore is conveyed into the Sag mill, which is similar to a ball mill in function. The SAG mill uses a light ball charge to grind the ore. SAG mills are larger in diameter than ball mills.

The mill has lifters built into the liner, like a ball mill. The charge of ore and balls rises and falls upon itself, and the harder portion grinds down the softer portion of the ore. Eventually, the ore exits the SAG mill and enters a ball mill for the final grind.

Depending on the ore, some operators do add a ball charge, but it will be less than the ball charge used in a ball mill, typically from eight to 20% of the charge.

The main advantage to the SAG mill is less wear from the light ball charge. Like any other mill, constant, regular maintenance is required.

There are both a vertical and a horizontal configurations out there, and there are arguments for the advantages of one over the other. The vertical mills seem to becoming more and more popular. Like their cousin mills, the rod and ball mills, the power requirements can be huge on the larger mills. The typical SAG mill will be shorter in length than a comparable ball mill.

Either way, the SAG mill should certainly be considered in any hard rock grinding circuit. Most, if not all, of the large mine operators have a SAG mill in the circuit. They are available in about any size you can think of, both vertical and horizontal, and they do save money by reducing the residence time in the ball mill.

Ball and Rod Mills

Ball and Rod mills have been around long enough to make stamp mills obsolete. Think of a rolling metal drum with a lot of different sized balls or rods being lifted and dropped on the ore being fed into the mill. Both of these mills are horizontal, and water is fed through the mill as is runs, and flushes ore out through the exit screens once it has reached the correct size. These mills are constant feed, and can grind a lot of ore.

The grind is determined by the ball charge and residence time in the mill. Keep in mind a rod mill is usually placed before a ball mill, and is used to produce a coarse grind. The ball mill then does the fine grind. Remember that the finer grinds are much more expensive to produce because of the energy requirements.

Ball mills are available in any size you want, clear up to the 15.5' X 25.5' 2250 HP Allis Chalmers giants you will encounter at the major mines. Hardinge, Marcy, Cascade, and Nordberg are some of the more common manufacturers. The large mills are usually started in the wee hours of the morning because they pull so much power off the grid. Starting them up guarantees a brownout in most areas.

Ball mills will accumulate gold. As the ore is ground, the gold particles, being malleable, will grind together and get larger, and some will actually go through the exit screen. When the ball mill goes down for a new liner, or an inspection, carefully collect that black, slimy stuff. To understand what you have, digest it in dilute nitric acid, HNO_3, rinse several times, and anneal in a roasting dish at $1,000°$ F to $1500°$ F for three minutes. The annealing will expel the nitrates, and you will see what gold is in it. After it digests, add some salt to the acid residue, and if you get a heavy white precipitate that turns purple in direct sunlight, it is most likely silver. If you have no chemicals, or chemistry skills, take it to your assay lab. A quick fire assay will give you an idea of the ratio of gold to silver, and obviously the amount of each present.

Rolls

A set of rolls can be quite useful on a lot of ores. Basically, they are two round, horizontal cylinders made of solid steel with axles through the center. One side is stationary, while the other is driven, and spring loaded. The axles are mounted on pillow block bearings, which are mounted to a frame. The tighter the spring, the finer they can crush. The drive side is usually powered by a chain drive, and the best models are made so that a pipe with the same inside diameter as the outside diameter of the roll can be slid over the roll and welded (tacked) in place. Lines of hard face can be welded on the outside of the pipe to prevent slipping. When the pipe has nearly worn through, it is simply replaced. The original roll surface is never used to actually crush rock.

Never use a belt drive on rolls. One hard rock, or a piece of tramp iron will lock up the rolls, and the belts will slip and jam the machine. When clearing jams, remember the spring may have tension on it, and will snap the rolls together. Many a mechanic and lab worker has lost fingers clearing jams.

Rolls come in about any diameter, usually from six inches for lab rolls to 16 inches for production rolls. Normally, the rolls are enclosed in a feed housing, and are fed a steady, constant stream of ore. Widths vary from six inches for lab rolls, to 18 inches or more for production rolls.

Once the spring tension or spring size has been established, and the feed rate set to produce the required grind, the operator is good to go. Overfeeding is the operator's worst enemy. Rolls are considered inexpensive compared to other methods of grinding, and are normally ran dry.

Iron Pyrite, FeS$_2$. Note the cubic cleavage.

A drum filter in operation at a mill in Honduras.

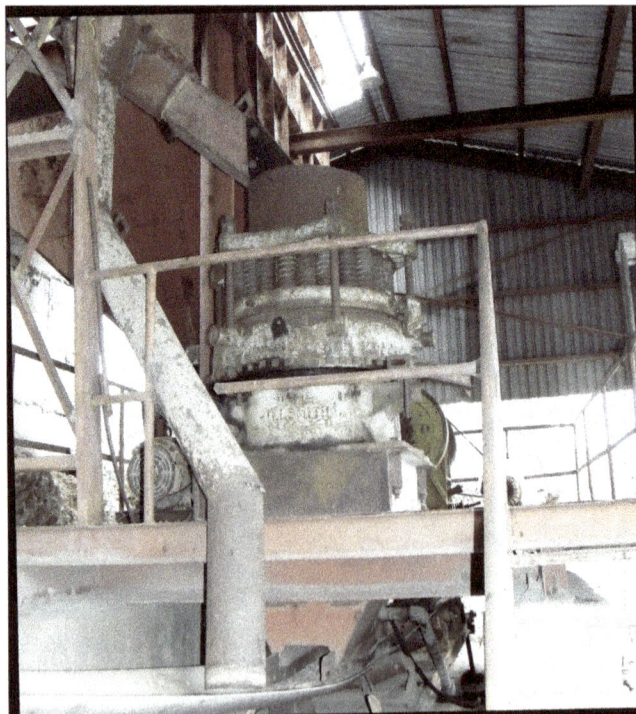
A cone crusher at an abandoned mine in Mexico.

A gravel plant set up and ready to go at an operation in Arizona. The feed hopper, jaw crusher and conveyor on the right were added on.

Typical hobby-sized trommel, with feed hopper. Great for pilot work on a placer.

Chapter Twelve

Test Equipment

Hobby Devices

There are hobby sizes of all equipment previously listed out there. As an example, Golden Manufacturing made a 1 ½ X 1 ½ jaw crusher years back. They were powered by a 1/3 horsepower electric motor. There's the Gold Cube, the K-GRS-1 sluice system, and more. The best place to find these devices is in trade publications, or Web searches by specific equipment types. They can be exceptionally useful *when you need to pilot a project without spending a fortune.* Most of these hobby machines are very well made, and easily capable of running a few tons. They are not toys.

Also keep in mind lab equipment. All labs have to have a crushing system, and there are small jaw crushers, 4X6 and up, pulverizers, and other types of equipment, such as small concentrating tables.

As a case in point, a few years back, some friends had obtained the coarse reject from a milling operation in Idaho. The operator had never ran screen fractions on the ore, had processed for the coarse gold, and left several ounces per ton in the oversize. The friends were given the material on the condition it was moved off the mill site. The material was moved to another property, where a quick analysis was done. A Bico disk pulverizer of the type normally used in assay labs was located, and a Gold Cube was purchased. The pulverizer would run about 200 lbs per hour. There were three people involved, so on days off and weekends they processed the material, recovering 99% percent of the gold. The equipment was dirt cheap compared to the recovery.

Sometimes, it is best to think out of the box. Fortunately, these folks were experienced in mining, and understood what was there, and the value of a simple fire assay. So, when you see some of this "mini" equipment, don't turn your nose up. If the small equipment will duplicate the mill or recovery circuit you have in mind, try it.

Think about the K-GRS-1 sluice system. It will process a five gallon bucket in five minutes or so. Got some placer samples that need processing? Seems that you could run a barrel in 15 minutes, and could recover the values, and calculate a recovery, again without breaking the bank. If your material has been ground, by all means, use the Gold Cube. Just be sure you have ground to the point of mineral liberation. If you process a hundred pounds, and get one pound of high grade concentrate, that's a concentration ratio of 100 to one. A good number to know. Assay the high grade con, and you can calculate a rough recovery. Always a good number to have.

There are also mini-trommels for the placer buffs out there. These little guys are basically miniatures of their big brothers, complete with sluice boxes and water tanks. Depending on the size of the trommel, the capacities can surprise you. They are easily capable of a pilot test. The key to all these devices is to head for the local farm supply, and grab up a couple 250 gallon stock tanks. This way, you have a little bit of a water reserve, and you can re-use the water. The first discharge into the tanks should be into a bucket or other removable container to catch the rocks and other heavy stuff. Any surfactants are kept in the tank, since you are recirculating the water. Screen the tanks so that you don't have a flotilla of ducks moving in, or livestock watering at the tank.

All this with "hobby" stuff. Keep these devices in mind. Another handy device is a small screen deck. Yeah, they make those, too. Interchangeable screen and all. Be aware of what's out there, and you might not have a big fancy mill, but you will have some of the numbers you need. Keep in mind that the Gold Cube and the K-GRS-1 will recover some really fine gold. Talk to the manufacturers about this fine gold recovery.

Metal Detectors

Metal detectors do have uses in mining, and at the large mines they are used to stop feed conveyors if tramp iron is detected. They can also trigger an electro magnet to pull the tramp iron off the conveyor.

The metal detectors you see running in the parks or on the beaches are actually quite useful on a placer project. As you noticed in the description of placer equipment, trommels use a barrel made of punch plate, with the (hopefully) appropriate sized holes for the nuggets and finer gold on the property. As the material is fed into the barrel, the material smaller than the punch plate holes passes through, and is processed through a sluice box or other recovery device.

The oversize gravel travels out of the end of the trommel, and is considered waste. Much to the chagrin of many an operator, large nuggets also go out the end of the trommel. Check your work. A metal detector is a cheap, easy way to do that. All metal detectors have an "all metal" mode, and that's pretty much all you need. Don't buy a $99.00 delight, and expect it to work well or for long. Try for a midrange machine, somewhere around $300 to $400. Whites, Fisher, Tesoro, Garret, and many others are out there. They all have excellent warranties and instructions.

The newer pulse induction machines are about additional depth on very small pieces of gold, and may or may not strike your fancy. They do not discriminate well, which really won't matter for your purposes. They are still fairly expensive, being new to the marketplace.

Your author has seen many a decent nugget come out of dry washer tails, and the oversize from placer operations. Don't let it be your nugget.

During mining, tunnel and shaft walls are obscured by dust, or to be more accurate, mud. The mud is mostly dust from blasting sticking to the moisture inside the diggings. A metal detector can help find metallic veins in the walls of tunnels, and the sides of the shaft. Cleaning the mud off and searching is considerably slower. Be very careful inside old diggings, they are deathtraps.

Chapter Thirteen

Process Circuit Design

Block Diagrams and Icons

By now you have obtained all the assay lab information you need. And, hopefully, having done pilot work, perhaps with hobby equipment, and you are ready to start laying out a circuit for your ore. Remember that the purpose of the exercise to this point is make you understand that each ore has different milling requirements. Hopefully, the ore has indicated that a moderate grind (if hard rock or ore pile) is all that is required.

It is always easiest to visualize the circuit with a diagram of some sort. This can be as simple as a block or box diagram, or one can use icons.

Here is an example of a basic circuit as a block diagram:

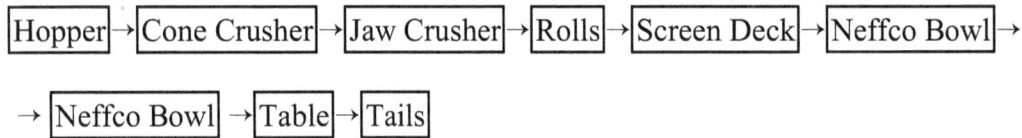

| Hopper | → | Cone Crusher | → | Jaw Crusher | → | Rolls | → | Screen Deck | → | Neffco Bowl | → |

→ | Neffco Bowl | → | Table | → | Tails |

The block diagram is effective, but if you are doing a presentation of sorts, icons are much easier for an audience to understand. Here are a few icons of the same circuit:

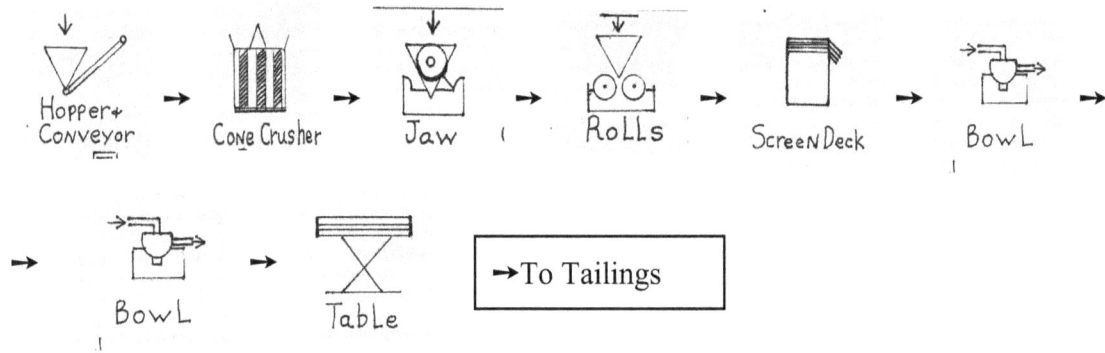

The choice is yours. There are a few icons on the next page. You can copy the page, or if you are a computer type, you can scan the page, copy the icon of choice, and place it where you desire. Or, you always have the option of drawing your own icons. If you are the engineering type, this is probably the way to go.

Equipment Icons

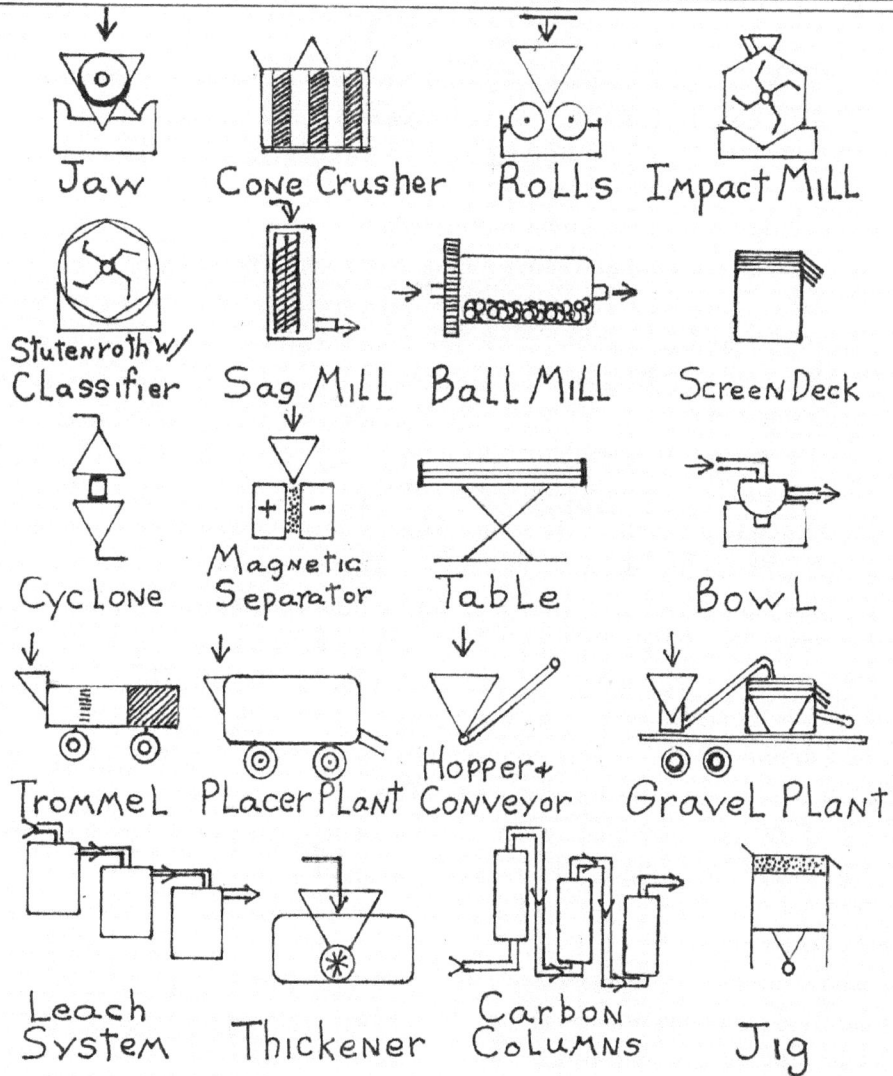

Jaw

Cone Crusher

Rolls

Impact Mill

Stutenroth w/ Classifier

Sag Mill

Ball Mill

Screen Deck

Cyclone

Magnetic Separator

Table

Bowl

Trommel

Placer Plant

Hopper + Conveyor

Gravel Plant

Leach System

Thickener

Carbon Columns

Jig

Hand drawn equipment icons.

Chapter Fourteen

Accumulating the Values

Concentration

This is what mining is. Concentrating values through various methods until at the end of the process, you have a bar of gold or whatever metal you are after. It doesn't matter how you go about it, cyaniding, placer mining, or gravity concentration, the end result is all the same.

As an example, look at a typical cyanide plant. The ore is crushed and placed on a pad. The cyanide will take the gold and other metals from say, 100,000 tons of ore into solution, say 10,000 gallons. That's the first step. Then the pregnant solution is put through carbon columns, where the gold is adsorbed onto 20 tons of carbon. That's the second step. The carbon is then stripped with hot caustic and stronger cyanide, probably 1,000 gallons. That's the third step. Using electrolysis, the gold and other metals are stripped from the solution onto steel wool. That's the fourth step. The steel wool is digested away with dilute sulfuric acid. That's the fifth step. The remaining sludge is collected, rinsed, and dried. The sludge is then fluxed, and smelted to produce a Dore' bar. That's the sixth step. The Dore' bar is shipped to a refiner, who refines the metal to final purity. That is the seventh, and final, step.

The values are the same at each step throughout the process. If the original 100,000 tons of ore assayed say, 0.080 OPT gold, the gold available would be 8,000 Troy ounces. Those ounces would be concentrated into the leach solution, 10,000 gallons, then concentrated again onto 20 tons of activated charcoal, or carbon. The sludge would weigh around half a ton, and when smelted produce about eight 75 Troy pound bars. Smelting will eliminate about 20 – 30 percent of the base metals, if correctly fluxed.

We also know that the average or typical recovery of a cyanide heap leach is 90%, give or take. So we would anticipate 7200 ounces recovered.

The point is that the *volume* of the gold bearing material is greatly reduced. A typical placer recovery may use different types of equipment, but the process is still a concentrating process. Just by running the material through a grizzly prior to feeding the undersize to a wash plant, you are still concentrating the values. You have removed the large rock from the feed. What percent of the feed would this be?

The diameter of the holes in the punch plate determine what will go through the nugget trap and through the sluice box again further concentrates the values. The oversize that goes out the end of the trommel again reduces the volume that goes all the way through the plant, allowing the sluice to gather more values, rather than waste capacity for oversize with no value.

At the end of the day, the sluice box will yield maybe 500 pounds of concentrate, or "cons" as they say. That material is again processed, usually with another gravimetric device, until the actual gold has been removed.

Hybrid, or Combination Circuits

Sometimes, a circuit is hybridized to enhance recovery. This happens way more often than not, especially in the case of free milling and complex ores that are blended by Mother Nature. For instance, a gravimetric circuit can be used to scalp the bulk of free gold and a leach circuit can be used to process the tailings from the gravimetric circuit to recover residual values in the tail. Another example would be the chemical recovery of PGM's from a gravimetric operation's tailings. Got sulfides? Split the circuit after the ball mill and autoclave or roast the ore, re-establish the chemistry and leach the gold with cyanide. Or put a flotation mill in to recover the sulfides.

Whatever method you end up using, perhaps now you are beginning to understand why running your ore through your Uncle Joe's mill isn't going to work. *Go where the ore takes you, if you expect to make a profit.* The trick is to use the KISS method, and do what you have to do to get a decent recovery. You'll never get it all, so don't spend a dollar chasing a dime.

A pair of Deister rougher tables used to concentrate scheelite, a tungsten ore.

Chapter Fifteen

Oxidation of Ores

Roasters

Got pyrite? Then you have a serious problem. Iron pyrite, FeS_2, is often alloyed with the gold in ore bodies. So, to recover the gold, you have to "oxidize" the ore.

Back in the good old days, the rotary kiln was the preferred and about the only method available. The ore was fed into one end of the revolving kiln, which was usually natural gas or propane fired. The heat would cause the pyrite to chemically separate, leaving behind iron, and of course, releasing sulfur dioxide into the atmosphere. This was a primitive method of roasting the ore, and quite expensive. Modern roasters are fed ground ore mixed with sulfur to sustain the reaction, and are still gas fired to the appropriate temperature.

Sulfur dioxide is scrubbed, and in some cases used to create sulfuric acid, which can be consumed on the mine site, or sold commercially.

Typically, these roasters are far and few between because of the high capital investment involved, and the permitting requirements.

On a small scale, Action Mining Services has invented a small roaster that is well worth checking out. Go to their website and look at the online catalog.

Autoclaves

Autoclaves come in all shapes and sizes, and normally they share two characteristics. Heat and pressure. The pressure cooker your Ma used to cook beans and can veggies is a form of autoclave. Medical facilities use autoclaves to sterilize instruments.

In mining, autoclaves come in two basic types, high pressure, high temperature, and low pressure, low temperature. The low pressure/temperature autoclaves aren't used nearly as much as the high temperature/pressure autoclaves are. The low temperature/pressure autoclaves typically use 120° F and 15 lbs per square inch. The high temperature/high pressure autoclaves typically work at 400° F and 400 lbs per square inch.

Autoclaves are essentially bombs. In fact, the test vessels used in labs are called "bombs" for a very good reason. They have been known to explode. Regularly. A high pressure/high temperature autoclave, can, and has, in fact, blown a 15 ton Lightning agitator through the roof of a building, and a few blocks down the street.

The high pressure/high temperature autoclaves are usually pressurized and heated with a boiler. The slurry within is acidified with sulfuric acid, H_2SO_4, and a certain percentage of sulfide is made available in the ore being oxidized in the autoclave. The available sulfide is typically 1.5% to 1.8%, and this is achieved by blending the ore down (or up to) to the required level. Some ores actually require the addition of a sulfide concentrate to maintain a stable reaction in the process. High pressure oxygen is also fed into the autoclave to assist the oxidation process.

Excessive sulfide can, and has, caused reactions in an autoclave to "runaway", or a pressure surge to occur. This creates an extremely dangerous situation, which must be dealt with quickly to prevent a rupture, and then a steam explosion.

Keep in mind the pumps and plumbing required to maintain the pressure. Ferralium alloys are the plumbing materials of choice due to the corrosion resistance of the alloy, and the Ferralium alloys are very expensive. The other problem is abrasion. The plumbing must be carefully monitored, or the slurry will abrade the Ferralium, which will weaken the wall strength of the plumbing, resulting in an explosion when the plumbing fails.

The high pressure/temperature autoclaves used in Nevada are typically 15' in diameter, and 70' or so long. The autoclaves have five agitated compartments that are separated by walls. The slurry overflows from compartment to compartment, the residence time within the autoclave controlled by the feed. Normally, this type of autoclave is constructed of 2" thick Titanium, and the entire interior is completely lined with acid resistant brick, and a lead based mortar is used to set the bricks. Periodic maintenance is required to maintain the brick interior.

All sulfides will oxidize in this environment, including cinnabar, orpiment, and realgar, The by-products are sulfur dioxide, and carbon dioxide, which are vented to a venturi scrubber where the gases are treated, and the mercury recovered.

Continuous monitoring of the sulfide in the feed stream is accomplished with the use of a Leco Carbon Sulfur Determinator.. Carbonate is monitored as well, as excess carbonate in the ore will evolve CO_2, which will cool the autoclave, and stop the oxidation of the sulfides. The oxidized discharge ore slurry must be "neutralized", cooled, and protective alkalinity established before the slurry is fed into a CIL or CIP cyanide circuit. The feed ore is usually ground to -100 mesh prior to being oxidized.

By now, you have come to understand that autoclave technology is very, very expensive, and can be quite dangerous if not operated properly.

Chemical Oxidation

As any lab person will tell you, ores can easily be chemically oxidized with a variety of different acids. The catch? The acids, even a technical grade, as opposed to reagent grade, are cost prohibitive.

Back in the good old days, various techniques were tried. At the mills in Cripple Creek, Colorado, and a few other places, high grade sulfo-tellurides were placed in a lead lined, sealed vat, and chlorine gas was injected. The purpose was to chloridize the high grade, creating chlorides, which were then processed through a cyanide plant. The values were swiftly extracted, as gold chloride is very soluble in cyanide. The safety problems with chlorine gas, and transportation issues stopped the practice. Another method used about the same time was to roast the ore with salt. The salt provided the necessary chlorine, however strict parameters had to be followed or a huge percentage of the gold would be lost in the process. Needless to say, with the current environmental regulations, you won't be using this method.

There is an enormous amount of information available on the older equipment and processes. The best place to start is with Sir T. K. Rose's "The Metallurgy of Gold", and "The Metallurgy of Silver". Both books have been re-printed and are available at different book sellers online. Google it.

A small operation in Arizona

Chapter Sixteen

Chemical Leaching

The Ore

The type of ore you have should determine the method of recovery. If you have an oxide ore, fairly low grade, say 0.5 OPT AU or less, chemistry might work for you, assuming the gold is in very small particles, and can be exposed to the chemistry without a lot of grinding. Of course, cyanide would be the chemistry of choice. The chemical has an excellent safety record, however permitting is tough. Containment vessels, pond liners, screening, fencing and monitoring wells are just some of the things you should expect. Heap or pad leaches require some acreage, and if properly constructed, can be used over and over.

A heap leach can produce gold for years if large chunks of ore are placed on the pad. It takes time for the chemistry to penetrate the rock, and dissolve the gold particles. The coarser the particles, the longer it takes. A combination of coarse gold, and large rocks on the pad means the pad could take years to complete the process.

Ultimately, the pad can be killed with a sodium hypochlorite rinse. The cyanide is destroyed, and the solutions are sent to the tailings pond, with a dose of Caro's acid (H_2SO_5) as a final kill. Note that Caro's acid is violently exothermic when being mixed. Don't even try to make this without a chemical engineer beside you. You will get hurt.

One of the tricks in cyanide leaching when the grade is too high is to dilute the values by blending the ore with barren country rock. This is fairly common in CIL or CIP circuits. Chemically there are all sorts of problems that high grade can cause, and the circuit simply can't recover the values due to suspension times, particle size, and so on. Tail assays start to climb, and the values that are lost are very difficult to recovery.

The Agitated Tank Leach

This method of leaching is actually quite common, and is often used in conjunction with heap, or pad leach processes. In those cases, the recovery medium is oftentimes added to the pulp, or slurry, hence the designations carbon in pulp (CIP) or carbon in leach (CIL). Typical percent solids is around 50%, and medium activity carbon (GRC6X8) in the amount of 20 grams per liter is added. The carbon loads as the slurry is agitated slowly, and is ultimately removed at the appropriate suspension time, to be stripped of values. Read more about stripping carbon in Chapter 15.

There a hundreds of arguments and debates about suspension times and agitation parameters, and there will no doubt, be debates until the end of time. If you are using a non-cyanide chemistry, and would like to learn a lot about agitation, go to www.asat.volant.org and check out the work done by Walt Lashley back in the day at the American Society For The Advancement of Technology. Walt wrote about this in a booklet called "Captured in Passing", Volume 6. In fact, if you contemplate non-cyanide chemistry at all, there are useful publications and lots of good information at this site.

Obviously, the information at ASAT is useful with any chemical leach, and is worth looking at.

Percolation

When doing a heap, or pad leach, if the solution can't percolate, or travel through the ore, there is no gold or low gold recoveries. Just like making coffee, the hot water has to contact the ground coffee to extract the beverage.

There are two primary causes for lack of percolation. The first one is clay. The clay will break down into smaller and smaller particles until percolation ceases. The second is simply over grinding the ore. The fines, with a small amount of clay will also stop percolation. Also, as the ore disintegrates into smaller and smaller particles, the ore itself may stop percolation.

The Column Leach Test

The standard test in the lab for percolation is a column leach test. A (hopefully) transparent plastic tube is stood on end, with the bottom set up with a filter, usually something like petro mat, or other coarse filtering material. The column is filled with the ore to be leached. The head ore grade is already known. The appropriate leach is applied at regular intervals from the top. The solution is extracted at the bottom and recovery, pH and all is monitored. Values in solution are normally tracked with an A-A. The progress of the leach is easily tracked in this manner. Obviously, if the solution is applied at the top of the column, and the total volume doesn't come thru the column once the ore is wet, there's a percolation problem. The milliliters of chemical leach solution is regulated, and measured, both in and out of the column.

The leach can even be put through mini carbon columns to test carbon extraction of the values during the test

A column leach can go on for a month or more, depending on grade, percolation, recoveries, and other factors.

The Bottle Roll

The bottle roll is an old, old assay lab standard. The roller is simply a framework with a motor geared to spin a driven rod, and an idler rod about six inches away supports the bottle. About thirty RPM is usually used. The two rods are rubber coated to prevent damaging the bottles. A piece of air compressor hose from a rock drill with a suitable inside diameter is fit snugly on the two rods. The wall thickness makes air compressor hose a good choice, but any rubber hose will work.

There are endless variations of this device. There are commercially made rollers, and homemade rollers. There are double decker, triple decker, and some that have been built to cover entire walls. They are easy to engineer.

Your ore is fire assayed for a head grade, and the appropriate type of ground sample is placed in a wide mouth, heavy bottle. Suppose you wanted to test a ⅛ minus sample to see how long it would take to leach the values in the ore. A 100 grams of ore, 300 milliliters of 6 lb per ton cyanide solution are placed in the bottle. Several tests are started simultaneously, with each test being removed from the roller at different times, say 12 hours, 24 hours, and 36 hours.

At the required time, the appropriate bottle is pulled off the roller, the ore rinsed well, dried, prepared and fire assayed. *Heads minus tails equals recovery*. The remaining cyanide can be titrated to calculate cyanide consumption, pH, and so on. Excess chemistry is used to insure there is enough chemical to do the job, hence the three to one ratio of solution to ore.

The bottles used are standard fare at lab supply houses, and are nice, stout bottles. The way they are made does not require a top, or stopper. A marker can be used to write the pertinent data on the bottles, then wiped off with acetone.

If you think about it, most any leach chemistry can be tested this way.

The Re-Useable Leach Pad

Yes, it can, and has been done. The thing to remember is that heavy equipment is going to be used on the pad to remove the tails, so get an engineer on board, and do it right. Your standard 4"concrete patio won't work.

Under the concrete, you will have to have a liner of some sort, such as Hypalon or EPDM rubber. Again, consult an engineer, the requirements vary from state to state, and change regularly. The pad will also have to be constructed with enough slope to drain fairly fast.

Your ore has to be clean for this operation as far as clay goes, because the plan is to bring in a gravel machine, and grind to ¼" minus, or maybe ½ " minus. Have a column test to determine the best grind, and percolation qualities. A bottle roll on your ore would be useful as well.

The ore would then be loaded on the pad with a layer drip tubes every two feet or so of depth, maybe topped off with a typical Rainbird type sprinkler when the pad is loaded. The cyanide solution is applied, drains, and is then pumped through three carbon pots, in the bottom, out the top is the rule. The chemistry is re-established in a collection vessel after the carbon pots, and pumped out again, until the pad is barren. A sodium hypochlorite wash for the barren material, and off to the tailings pile. Be careful to keep the hypochlorite rinse away from the leach liquor.

Normally, with 0.25 OPT gold, the cyanide would be set around 4-6 lbs per ton. One lb per ton of cyanide equals 500 PPM.

The Chemicals

There are other types of non-cyanide chemistry out there, such as thiourea, the bromines, (Geobrom 5500), urea, chlorine, and believe it or not, plain old hot water. Geobrom, one of the more popular products , has been re-packaged and sold for years under various names. It actually works quite well on some ores, usually oxide. It was manufactured by Great Lakes Chemical as an algaecide for cooling towers. If you tech types really want to know, it is 1,3, dibromo, 5,5, dimethylhydantoin. It does tend to evolve some nasty gases at very high or very low pH.

Once you get away from mainstream chemistry, you should plan on doing your own tests. Leach tests at the bench top level are pretty straightforward, and do not take a lot of expensive equipment. Hotplate stirrers, pH meters, thermometers, a vacuum filtration system, and a good ventilation system, for starters. A fume hood is a very good idea. You will need access to a chemical supply house, and have some credentials to be able to purchase chemicals. Homeland Security monitors all chemical sales that involve precursors for the manufacture of assorted types of dope, such as meth, speed, etc. Or explosives. Sadly, the mineral acids you will need are used in the manufacture of all of the above. Trust me, they will come see you, or at the very least, check you out thoroughly.

Your chemical supply house will obviously be concerned about their liability if you accidentally detonate the grade school next door, or the vapors kill off the flowers, pets, and children next door. That means you need to have a suitable place in the correct zoning to proceed, and some background knowledge of chemistry and chemical procedures.

This pilot plant outside Wells, NV was for cyanide CIL, and flotation.

Chapter Seventeen

Leach Recovery Systems - Carbon

Activated Charcoal

Activated charcoal, or for the purposes of this book, "carbon," is made from coconut shells. The shells are ground, and burned in a reducing atmosphere, so in effect, you get charcoal instead of ash.

Activated charcoal has hundreds of uses in the water treatment area, and can even be silver impregnated as a bactericide in wastewater treatment. The use of coconut shell carbon in the mining industry was implemented in the early to mid 20th century.

From a chemical viewpoint, the covalent charge of the carbon particle is stronger than the weak ionic bond of a gold molecule to cyanide. The gold particle (among others) is adsorbed onto the carbon particle.

Back in the day, many different types of charcoal were tried for recovering gold, but they were not hard enough to withstand the action of an agitated leach. All carbon will attrit, or abrade to fines in the process. Therein lies the rub. The finer the carbon, the heavier it loads with values, such as gold and silver as well as trash elements such as iron, copper, arsenic, mercury, and so on. The trash elements consume the space that could be used for precious metals.

Carbon fines are almost impossible to contain, and will migrate throughout any leach system. New carbon is placed in a tank full of water, and slowly agitated to wear away the loose carbon fines. This process is called attrition. After attrition, the carbon is washed on a screen and is ready for use. After attrition, the majority of the fine carbon will have been removed, but fine carbon will be shed in the process. This is considered an acceptable loss by the operator of the system.

Note that carbon comes in different sizes, and the size is usually preceded by the letters "GRC", indicating Gold Recovery Carbon. Sizes typically run from 6X8 (coarse) to 8X12, (fine), and carbon can be bought ground to any mesh size.

If you look at a particle of carbon under a strong microscope, the carbon is porous, allowing the carbon to load way beyond the surface capacity of the particle.

Most carbon is produced overseas, with Sri Lankan carbon being one of the more common types available. The Philippine Islands produced a large amount of carbon as well. There have been endless debates over the years about who has the best carbon. The best way to find out it to get samples of each kind and test it yourself.

Carbon can come with in three different activity levels, low, medium and high. Normally, medium activity is used in mining as it will typically load to around 600 OPT of precious metals. Low activity will load to about 300 OPT, and high activity carbon can load to as high as 1,000 OPT. High activity carbon will not strip beyond 600 OPT in the normal processes, and must be destroyed to recover the remaining values.

Normally, carbon can be loaded and stripped five or six times. Eventually, the carbon particles wear flat ("platelets") and become so fouled that they can no longer load values. This carbon is normally destroyed. There are residual values in this carbon, usually from 6 OPT to 20 OPT, so destruction of fouled carbon can be a profitable enterprise. Fouled carbon will also have the undesirable elements typical to the process, such as iron, mercury, arsenic, copper, and so forth. Take care not to discharge such metal oxides and mercury vapors. Also, carbon destruction is a thermal process, so the discharge of CO and CO_2 is an issue. Permit accordingly if you do this.

Typically, a heap leach will recover loaded cyanide solution in the "pregnant" pond. The pregnant solution is pumped through carbon pots, or towers that are half full of carbon. In the bottom, out the top. Usually, there are three pots or towers, smaller pots being in a cascading arrangement. The first pot is higher than the second, and the second is higher than the third. The pots are usually quite distinct at heap leach sites.

The pregnant solution is circulated through the pots or towers until the values are acceptable to the operator. The chemistry is re-established (pH, free cyanide, and so forth) and is pumped back to the leach pad to be loaded again and again.

The operator chooses the configuration of the system, and there are always debates about the best setup. Canadians have one way, the Australians another, and so on. The Australians patented a system that uses packed carbon columns, and of course the argument was that the stream of solution would cause channeling in the carbon, and result in uneven loading of the carbon. The debate goes on.

In a CIL (Carbon in Leach) or CIP (Carbon In Pulp) leach system, the carbon is added directly to the slurry of oxide ore. GRC 6X8 is the most common size. The carbon travels up the leach tanks with counter current decantation, and the residence time is established by metallurgical test work. At the specified time, the loaded carbon is screened from the slurry, rinsed, and sent to be stripped of the values it contains.

Stripping Carbon

There are two methods of stripping carbon. The first, and oldest, is the Zadra, or atmospheric (no pressure) strip. The second is the pressure strip, in which the entire process is enclosed, and placed under pressure. Typically, the pressure for the pressure strip is generated by a boiler, and is around 15 PSI.. The idea is to chemically treat the carbon with a stronger cyanide solution, and electrowin the solution from the strip with steel wool. As always, protective alkalinity is set with sodium hydroxide, NAOH. If the initial leach was two to four pounds per ton of cyanide, the strip will be at least 10 pounds per ton, and depending on the operator can be as high as 35 pounds per ton.

The Zadra strip is non-pressurized, and has been used for many years. The pressure strip is considered faster, and more efficient. The best strips are accomplished by "shocking" the carbon with hot, strong chemistry. This approach typically is used in the Zadra strip, and can exceed the pressure strip in gold recovery.

In either case, the values are deposited on steel wool in the electro winning cell, usually running two to four volts of DC current. The amps (throw) are regulated by the amount of steel wool, chemistry, distance from the cathode, and so forth

The steel wool is digested in dilute sulfuric acid. The residue is rinsed well on a tight Whatman filter paper, dried, fluxed, and smelted. The metal product of the smelt is called

Dore', which is a mixture of the metals captured through the process, and is by no means a pure product. The Dore' is sold to a refiner that charges a small percentage to refine the Dore' to pure metal products, such as gold and silver. If a certain percentage of an undesirable element is present, such as iron, it is considered a penalty element, and a fee is assessed the seller for the extra cost of separating the offending element. Hence the term, "penalty element".

Keep in mind any element can be a penalty element. A good example is platinum. Platinum in small quantities is pretty common at large leach operations. You have 1.5% in your Dore', and depending on the refiner, it could take 2% to get a paycheck on that element. The element may be valuable, yet there is just enough present to increase the refiner's costs, so the platinum can be considered a penalty element. Your refiner will provide a schedule of fees, make sure you know what they are.

Some refiners will also have a "primary metal" schedule, where you decide what your primary metal is, such as gold, and the refiner will pay only on that element, but usually will pay much quicker. If you are producing a gold and silver Dore', and select gold as the primary metal, you won't get paid for silver. Usually, on a primary metal arrangement, the percentage of penalty elements may be higher, saving you the extra fees. Work this all out with your refiner, way in advance.

Recycling The Carbon

After the strip, the carbon should be put across a bar screen to remove the flattened particles, or "platelets". The carbon is then put through a rotary kiln at 1000° F and quenched in cold water. This "popcorn" effect will pop the outer layer of the carbon off. This removes the soluble silicates and other trash, exposing a fresh surface and unblocking the pores of the carbon. The carbon is then rinsed, screened to remove any fines, and returned to the leach circuit, or carbon pots to be loaded again.

The carbon will load less and less with each successive cycle, until it has the final strip, and is considered waste, or "fouled" carbon. This scrap carbon can then be ashed, or destroyed to recover the remaining values.

Ashing Carbon

This process is becoming fairly common, and there are some advantages to be taken by the astute operator. As previously mentioned, the strip circuit can be expensive to operate and build. There are obvious risks associated with cyanide and hot caustic solutions. Ashing or destroying the carbon can be used to eliminate the strip process, regeneration with all the necessary equipment and manpower, and the handling of the carbon.

Assuming that the operator can find a responsible ashing or carbon destruction facility, the operator should contemplate loading high activity carbon to the maximum. The operator should verify that the carbon purchased is at a competitive price, and that the carbon is, in fact, high activity. The initial cost of the carbon is a factor in the calculation of this process.

The operator would then load the carbon to the maximum, which should be near 1,000 OPT of precious metals. The carbon can be sold based on an assay. Carbon is very difficult to assay in the fire assay, and the assay of the carbon by x-ray diffraction should be contemplated, or sought. The carbon is then sold for a large percentage of the agreed upon values. Quick cash. No waiting, and no penalty elements. That is all assumed by the buyer.

Good operators understand that income is income, whether it be a check from a refiner or a carbon destruction facility. The operator should understand that the Dore' bars are simply a means to an end, and not allow the ego to get involved.

There is at least one commercially available process available. This was invented by your author back in the 90's and is still in operation today. Your author understood that a small to medium operator would save considerable startup costs by not installing a strip and regeneration circuit. The process also provided a means of recycling fouled carbon for any operation. Many operators would simply bury mercury laden fouled carbon on the mine site, rather than find a legal method of disposal.

Another advantage of ashing carbon is the fact that the product of the process is an oxide. All the elements are reverted to their metallic state for easy recovery. The exception is mercury, which is vaporized during the process, and requires treatment of the vapor stream to prevent atmospheric discharge.

Column leach test in progress.

Chapter Eighteen

Other Leach Recovery Systems

Resins

Many chemical leach operators have pondered the use of resin beads as a recovery media as opposed to carbon. There is no doubt that resins will strip precious metals from solution. Keep in mind that the resins are a solid sphere, and to create surface area they are very small. Resin beads are very difficult to contain due to their size. A small hole in a fine screen will allow contamination of the entire circuit.

Your author has done loading tests on 25 or so different resins over the years. The most common water treatment resins seemed to load the best and the heaviest. Resins will not fire assay. Chemically strip the resins, and determine the values with an Atomic Absorption Spectrophotometer.

They also offer some chemical challenges when it comes to the strip process. Granted, Aqua Regia will strip precious metals from most any substance. For this process to succeed, the containment and handling of the resins can be quite a challenge. To attempt to strip resins on a commercial scale would be mind numbing.

Some resin manufactures recommend stripping with hot Hydrochloric acid, however this doesn't eliminate the containment problems, and rarely is the strip thorough.

Keep in mind that a lot of the "substitute" chemistry used to eliminate cyanide will load carbon, so keep your options open, and test your leach chemistry with carbon.

Zinc Precipitation

Zinc in one form or another was around for many years, and has been used with about every chemical leach known to man. Zinc boxes, zinc dust, and mossy zinc all work the same way. The zinc simply replaces the precious metal ions with zinc. The catch is that sooner or later your leach solution will have less and less chemical available as more and more zinc goes into solution.

The "zinc box" method was popular with silver miners using cyanide for many years. At that time zinc was dirt cheap, carbon was not. So the silver was precipitated with zinc as opposed to loading on carbon.

Direct Electro-Winning Precious Metals

Some operators will claim this cannot be done with cyanide, and they are wrong. There are a lot of factors that have come together for this to work. It is especially useful for low and medium grade silver ores. The cyanide is set up fairly strong, usually around 10 Lb. per ton. The pH is set to 9.5 to 10.5. The solution is routed through an electro-winning cell, and following the electromotive series, the appropriate voltage and throw are set for the element being recovered. Typical cells have stainless steel plates, and the silver will weld to the stainless plates. A little lead sugar (lead acetate) will combine with the silver, and create big, fluffy precipitate that is easy to recover.

The precipitate is collected on a fairly tight filter paper, rinsed, dried, fluxed and smelted to Dore'. The lead acetate will create a small amount of lead contamination. When you flux, add a small amount of sodium nitrate to your flux for the extra oxygen, and fire an extra half hour. What little lead there is will slag off. Keep in mind silver is self-reducing in a smelt. If your slag is off white, it still has silver in it. Re-fire the slag.

Chemically, most any solution containing precious metals can be stripped by electrolysis. The main points are to use a full wave rectifier, not a battery charger, and to follow the electromotive series for the element you are after.

Flotation

Flotation mills (or just float mills) were normally used in ores with a high sulfide content, or some sort of offending penalty element present, such as arsenic. The chemistry could be altered to float or depress most any element.

Float cells came in banks, with any where from four to eight cells in a bank. Each cell had an agitator that would blend the appropriate reagents to create a froth, like soap suds. The fine particles of ore would adhere to the bubbles, which overflowed into a trough. The trough carried the particles away for rinsing (sometimes), and to a filter press to remove excess chemistry.

The ore was finely ground, at least -100, and fed to a tank where the reagents were added, and then fed to the float cells. Recoveries weren't sky high, and all the penalty elements led to the demise of flotation. The main problem was that flotation concentrates were smelted. The EPA made smelting go away by tightening (restricting) the emissions on discharge permits.

A lot of float concentrate went to Canada for a period of time, and then the Canadian smelters also went away. Flotation got the gold, but it also captured a lot of elements that couldn't be smelted. This was pretty much the way the old silver smelters went. The lead in the ore, which was usually a lead silver sulfide, Galena, was contaminating large areas, and then there were the issues with sulfur dioxide, SO_2.

Note that China, and other countries continue smelting and discharging and contaminating. There are new technologies out there that can be of considerable assistance in smelting, but for the most part, they are considered cost prohibitive at this point.

Chapter Nineteen

The Universal Solvent

Water

Cool, clear water. Typically, a ton of mine run ore will need around 40 to 50 gallons of water *just to wet the ore.* Be ready for that. In the desert, it takes even more water. Even running bulk filtration systems, you will use a lot of water. A ton of water is 269 gallons or so, depending on the specific gravity of the water.

This is a major consideration anywhere in the world you may be setting up or running a milling or placer operation. For milling with chemistry, the water you use in your circuit *must be drinking water quality.* Whatever the water source, be it a well, a stream, whatever, the water must not have a total hardness over 200 PPM or so, and a pH of 6.5 to 7.5. If it does, just like drinking water, you'll have to treat it. The main elements that contribute to hard water are calcium, sodium, potassium, soluble silicates, and iron. There can also be other metals in the water, such as lead, arsenic, and so on. Test your water. Water analysis is pretty common these days, and fairly inexpensive. All those hardness elements will displace gold in solution, can cause gold to precipitate from solution, and so forth.

Placer mining? Well, the operative words here are *"clear water".* If your feed water is a stream or river, don't feed your plant muddy water. You may have to clean the water intake to your plant. Run the plant with the water available, and try running a large sample in a test plant with clear water. Observe any differences in recovery. It can easily be worth the effort to clean up the water. It is illegal to discharge muddy water in any case, so put in stilling ponds or tanks, and find a surfactant (breaks surface tension) and a flocculent (filtering or settling aid). Both are cheap, and fairly common. Small placer operators have been known to use Jet Dri, Lysol, 409, and other super cleaners for a surfactant. As long as you are using a closed system, it's easy to do. Very small quantities are all that's required. Sodium Hydroxide as a fine gold depressant will probably require permitting, and security netting to keep the wildlife and ducks out. Some operators will simply throw a carpet mill, or miner's moss at the end of the circuit to catch the fines, and forget the chemical depressant.

Make sure your equipment isn't dripping motor oil in the stream or river. This was one of the problems in California that shut down dredging. It is called "contaminating the resource".

Flocculent

A flocculent, typically, will be one of three kinds. Anionic, cationic, or non-anionic. The anionic and cationic are based on the charge of the suspended particle (colloid). The particles are drawn together, forming "flocs" which become heavy enough to quickly settle to the bottom of the tank, or container. It is impressive to watch how fast the particles settle, and how clear the solution is when using the right product.

The non anionic flocculent has no charge, and is basically microscopic sticky strings that collect particles, become heavy, and sink to the bottom of the container. It does the same job, just differently.

For many years, the flocculent of choice was those made by American Cyanamid. They later became just Cyanamid, and then were spun off and are now called Cytec. They still list mining chemicals on their website. The product line back in the day was called Superfloc. There are many different types of flocculent available. Cytec is not the only manufacturer, most major chemical companies manufacture or distribute "settling aids". They are used extensively in water and sewage treatment. According to www.thomasnet.com, there are 77 manufacturers in business. Or try SupplyMine, as listed with www.infomine.com. You can buy flocculent from Amazon, however you will have to buy a quart or so just to see if it works. Typically, these are pool or pond clarification chemicals.

The best method is to contact a supply house, such as HyChem, and talk with a salesman about your application. It would be helpful to know the pH of the solution you are trying to clarify. The salesman can provide a range of samples, and instructions on testing the product they supply. The tests are fairly easy, and don't require a degree in chemistry. Keep in mind these products can be a powder, or a liquid that you dilute out to a specific concentration, or even a solid roll of material that you just let your slurry run over. As the slurry runs over the roll, enough flocculent will dissolve to clarify your solution.

Remember that coagulants and flocculents are used in conjunction with thickeners, drum filters, belt filters, filter presses and so forth. A simple setup would be a pachuca tank, which has a cone bottom. A slurry would be fed in the tank, the flocculent would be injected into the feed line, and after settling, a slurry pump would pump the settled material from the tank. Note that if the slurry is settled, then agitated, it will have to have another dose of flocculent to settle again.

For large volumes, a thickener such as a Dorr thickener is used. Dorr thickeners have been around for over a hundred years, and were even made of wood at one time. Some miners have welded up "mini-thickeners" based on the Dorr machines. If further dewatering of the slurry is required, a belt filter or similar device is used on the slurry from the thickener.

Mechanical Filters

Mechanical filters, such as swimming pool filters, are useable on a small scale, however they do not remove colloids in suspension. The colloids are simply too fine. A flocculent causes the colloids to clump, so they will filter easily. Think about a belt filter.

Nalco Water Book

A must have book for water information is the Nalco Water Handbook. Nalco manufactures water treatment chemicals for about every application, including mining. The book explains everything you want to know about water, and water treatment. These days, you can download the book as a PDF file for free. Google it, and then download. You can even get a copy for your Kindle, if you wish. Nalco has offices in most states, and has mining chemical offices in some states, such as Colorado. Check out the website.

Chapter Twenty

Contractors

Contract Mining Companies

Suppose you've found it, but can't or don't want to mine it. Hard rock mining is not a cheap thing to do these days. The capital cost of the equipment, explosive permits, and numerous other financial considerations would dictate that you should find a contract mining company. If you are on a typical mining claim, it can take months to be permitted by the appropriate agency, such as the BLM, so plan on that, as well.

There are companies out there that range from small, clear up to a million tons a day. A quick Google search will bring up Redpath, Ledcor, and many others. Remember that most of these companies have a tonnage requirement, and to them, 10,000 tons is just too small. If you also check the trade publications such as the International California Mining Journal, you will find other leads.

Always check a site that is in operation if you can. Also think about and check safety records, MSHA records, and references if you can get them. Don't ask if they have insurance, ask to see policies. Check with the State Mining Board for pertinent information. Make sure the IRS doesn't have them under investigation. Do they have the appropriate business license?

Any legitimate company will be doing Federal withholding, unemployment compensation, and most importantly, workmen's compensation. Mining is a dangerous business with plenty of liability to go around. If someone is seriously injured, or killed in an industrial accident, what will your liability be? Make sure you got it right, or you will be paying the rest of your life.

Some things to think about. Sampling. Will the contractor provide drill cuttings for assay? Does the contractor have an assay company they work with? How far will they transport your ore? Any contractor worth their salt will insist on an assay program to keep track of the ore body. This is also highly beneficial for you, those numbers will tell you where you are, and keep your contractor headed in the right direction.

Once you have done your homework, and have worked up a satisfactory contract, an engineer will have already evaluated your property. The engineer will be a tremendous source of information. *Listen to what he says.*

Once the contractor begins, it won't take long to see what appears to be a surprisingly small pile of ore. Fifty thousand tons seems to be a lot, but the pile won't be as big as you think.

Security

Normally, the ore is transported to a prepared site and stacked. If you have any visible free gold, be prepared for your pile to disappear. This happens a lot.

A client of your author discovered a huge chunk of "sugar quartz" studded with gold. This slab was about 30' thick, 60' long, and 50' or so deep. The client found this on a Friday, and the following weekend was a three day weekend for the 4th of July. On Tuesday after the weekend, the client went to start mining the slab, except it was gone. It had been mined and trucked from the site in one weekend. The missing ore was found a month later in a secure warehouse 250 miles away. Four guys went to the penitentiary for that one.

In Mexico, Central and South America, the stories abound of mining companies doing exploration, and shooting a round in a rock face. Returning the next morning, every pebble will have been hauled away by the indigenous people. The residents figure that the rock must be worth something, since it had been drilled and blasted by a mining company. So don't be surprised at this type of behavior.

Use your head. Build fences, or have a watchman on site. Any mining operation will have the casual lookers, and every one of them will haul off specimens. If you have really good 200 OPT gold ore, better have some armed guards on site. In Central and South America you'll need some serious firepower. Hire the local military, but don't be surprised if they hijack the goods and you. Good luck if you are overseas.

Contract Drilling

Same deal as contract mining. There's lots of contract drillers out there. They will probably require your property to be surveyed, and access right of ways verified. Normally, drillers collect drill cuttings as part of the package. Make sure they do. You might have to provide sample bags, but it is well worth it. Also make sure they properly plug and mark all drill holes so you can easily find them if necessary.

Modern drills are reverse circulation, and the drill cuttings are flushed up and out as the drill works. Again, these companies have rules to follow, and require a minimum footage before they stand the expense of moving the drill rig(s) on your property. A visit to the company's office will answer most of you questions.

You might have to go to a water well driller in your area, if no specific mining drillers are around. All active mining areas have active drillers in the area. Think about Elko, Nevada, and all the mining activity in the area. A local water well driller can probably steer you to a driller in your area, if nothing else.

Contract Milling

Or "custom" milling. This is the only place in the universe where gold is "lost to solution". It is sad but true that to the day this was written, your author has never encountered a custom milling operation that was truly legitimate. The usual process is that a group of people, usually dishonest, will build or lease a mill, and advertise their services. Normally, the mill lease and or equipment are purchased with investor's funds, not the principles, or operators money.

The operators assign themselves some title, such as "Chief Extractive Metallurgist" or "Mill Recovery Superintendent", and they will know all the buzz words. They will proudly proclaim to have a "newly discovered method of chemical extraction" which may, or may not include a black box to provide some unspecified function, usually dealing with incredible voltages. At this point, you should have started to sense the oily presence of a snake oil salesman, and hopefully, you will not walk away, you will run.

The other method is to assemble a milling operation to feed a gravimetric circuit with bowls, tables or whatever. Same deal. Damn! No gold in the forty 40 tons they ran for you, must have been a bad run from the mine, they'll say. Reality? They simply put a "T" in a PVC feed line, and scalp the gold before it reaches the table, bowls, or whatever. If you don't understand milling, and recognize a nugget trap, you're screwed before you start. And to think, you paid for each ton to be processed, on top of that.

This is just another form of theft. These operations have popped up all over Nevada and other states from time to time. Usually, by the time the authorities arrive, these people are off to the next location, with a new name, and a new LLC, doing it again. Usually, the main trouble they get into is for failure to permit the chemistry they are using.

The main problem with all this is that the victim rarely understands what happened, and has insufficient proof of a loss. Remember all that stuff about assays? You need to know what you have, and understand the terminology. These guys will rarely, if ever, have a lab, and if you ask, they'll tell you they don't need one. Yes, time to run again.

Your author has been hired many times over the years to investigate this type of operation, and some of the circumstances were incredibly embarrassing for the victim. All the warning signs were there, the victims just didn't pay attention.

Sampling crew attempting to locate values on a hillside.

Chapter Twenty One

Pilot Testing

The Pilot Mill

The purpose behind a pilot mill is to duplicate the mill you anticipate building, and establish and verify the recovery of values from your ore. An assay can tell you a lot, but not how well your anticipated circuit will work. You have to establish the actual parameters of the mill, and then upscale to a full sized project. The pilot operation will also prevent some nasty surprises, so it's not a good idea to have dreams of sugar plum fairies and gold bars. Better to find out as much as you can, and a pilot operation will allow you to take your ore all the way to a gold bar. Small, perhaps, but a gold bar, nonetheless.

We will look at each of our projects as we go, and figure out a pilot circuit for each. Understand that each pilot operation is a huge assay. You will be verifying all the information you have so far. *It is not etched in stone what type of circuit you have to have!*
Just because it's hard rock ore doesn't mean the ore has to be leached. Find what will make a profit for you.

The Hard Rock Ore Pile

The first order of business is to review your assay report, and hopefully, you have more than one, now understanding the value of assays. As you can see, you have just over 5.00 OPT gold. We know that this is mine run ore, so we will assume, for the purposes of this exercise that the maximum rock size is six inches.

First, let's review the screen fractions report on the next page:

Acme Assay Company
123 Golden Way
Somewhere, NV 89706

Assay Report - Gold Only - Screen Fractions #3

John Smith Mine   **123 Easy Way Rd.**   **Gold Mine, AZ**

Sample ID:	Sample Name	Au, Oz/Ton	Ag, Oz/Ton
1	-10 to +12	0.001	N/A
2	-12 to +15	TR	N/A
3	-15 to +20	0.002	N/A
4	-20 to +30	0.001	N/A
5	-30 to +40	0.089	N/A
6	-40 to +50	0.150	N/A
7	-50 to +60	0.250	N/A
8	-60to +70	1.400	N/A
9	-70 to +80	1.350	N/A
10	-80 to +90	1.580	N/A
11	-90 to +100	0.200	N/A
12	-100 to +110	0.08	N/A
13	-110 to -120	0.002	N/A
14	-120 to +130	0.001	N/A
15	-130 to +140	0.001	N/A
16	-140 to +150	0.002	N/A
17	-150 to +160	0.001	N/A
18	-160 to +170	TR	N/A
19	-170 to +180	0.001	N/A
20	-180 to +190	TR	N/A
21	-190 to +200	0.001	N/A
22	-200 to -250	0.001	N/A
23	-250 to +300	TR	N/A
24	-300 to +350	TR	N/A
25	-350 to -400	NG	N/A
26	-400	NG	N/A
Note:	NG = No Gold TR = Trace		

Date: 2/14/2015   Assayer: Leroy Jones

Now you have a choice or two, which will depend on your situation. Do you buy a jaw crusher, buy or lease a gravel plant, or talk to your lab and hire them to crush for you? The lab is probably out of consideration due to the large volume of crushing you will do.

If you have 50,000 tons, you probably want to rent a gravel plant and a front end loader. Set everything up, and plan on running the plant for three or four days, maybe a week. Work your way down the ore pile, and crush a few hundred tons at four different spots, moving the plant far enough from each run that you won't contaminate the other runs.

At the end of this, you will have four different piles of eighth inch minus ore, or even quarter inch minus ore. Now we look at our screen fraction reports, and plan our grind for milling. If our ore pile is showing screen fractions indicated on report #1, we know that we must grind to -100 mesh to liberate the values.

Suppose we have found an impact mill with a classifying air cyclone at an equipment dealer, and we can rent it or lease it for a couple of weeks. The impact is rated at *10 tons per eight hour day*. Your gravel plant has left four piles of 200 tons each. The math to calculate the volume of a pile is pretty standard, all you need to know is how much a cubic foot of your material weighs. Have your carpenter or welder make a box that holds exactly one cubic foot. Fill it, scrape it level, weigh the box and material, deduct the weight of the box, and there you have it. We know there are 27 cubic feet in a yard. Do the math.

We will assume, for the purpose of this exercise, that there are 3500 lbs per yard, dry weight. So you have 1.75 tons per yard, and just a little over 114 yards per pile. This may seem ridiculous, but you need to have the numbers for your final recovery calculations. This is what will tell you if this is a profitable operation or not.

You know that -100 mesh is a fairly coarse particle, and a good gravimetric circuit will recover almost all of the values, so you can get a couple Neffco bowls with a feeder, and a good 4 X 8 concentrating table. Grab a couple 500 gallon stock tanks, and some common flocculent. Neffco bowls are rated at four yards per hour, but that is actually a little high. Set the bowls up so that the first bowl's outflow feeds the second (surge) bowl. The outflow from the second bowl feeds the table as a safety measure. A short hillside is a good place to set the circuit so that you can use gravity to do most of the work. Later, you will use the table to clean up the concentrate from the bowls.

The first stock tank will have to have an auger mechanism to remove the waste, or have a large forklift on hand and just dump the tank when it's half full or so. The first tank drains into the second tank, where the flocculent would be applied. A floater in the middle of the tank would hold the hose that would be used to pump the water back to the bowls. If you have a handy source of water, use the tanks for settling ponds so you don't discharge muddy water back into a stream or pond. You'll choke the fish with muddy water.

A feed hopper and belt are good to have to feed the system. A constant steady feed is necessary to prevent surges to the bowls. The air classifier can feed directly into the feed hopper, or the feed hopper can be fed with a small loader.

If you feed a slurry to the bowls using PVC or other plastic pipe, "T" in a short piece of pipe ahead of the bowls and use it as a nugget trap. Make sure it is threaded so you can remove it. You can probably catch 90% of the gold right in the trap.

The feeder for Neffco bowls is basically a rain gutter. As the water from the second tank is pumped into the end of the feeder, dry, classified ore is feed into the feeder, and is a slurry when it reaches the bowl. The bowls will spin, and the lighter material will move downstream to the second bowl, and overflow will routed downstream to the header box on the table.

Make sure your table is on a level concrete pad. The bowls are usually bolted on a stout angle iron frame. Once you have the table set up, the operation will pretty much just run, until you stop. Let the bowls wash out, the table run out of feed, and then shut off the water and empty the bowls. A large gold pan is just right for this.

Usually, a five gallon bucket is used to capture each cut on the table. You will be especially interested in the first cut in the event there is a problem with the bowls. *Never, ever discard any cut off the table, or the tails, until they have been assayed.*

This circuit is simple, easy to use, easy to set up, and in your author's experience, has always recovered over 90% of the values described in the scenario above.

You can only learn when to empty the bowls by experience. Usually, they are checked after the first hour, and if there are no problems, then they are checked after the second hour, and so on. Flush them into a pan or bucket with a garden hose after pulling the drain plug. It only takes a few minutes, so it's always wise to check your work.

The concentrate you collect from both the bowls and table will consist of magnetite, hematite, gold, and any other dense particles. If you used your head, and you picked a really good table, you should have no trouble changing to fresh buckets at the table, and running the concentrate to clean it up. Or, if you wish, you could purchase a lab sized refiner's table, and you'll be ready to smelt to Dore'.

Any decent table will separate the iron and crud from your gold. Don't attempt to smelt your gold unless it's really nice and clean, and at least 70% gold. The refiner's flux, or "Chapman" flux in the book "How to Smelt Your Gold and Silver"can slag off some iron, but not all if the iron is in preponderance.

You can estimate your recovery simply by knowing how much ore you have processed. You should recover at least 4.5 ounces of gold per ton. Your 200 ton pile should yield 900 ounces of gold. At the time this was written, gold was hovering at $1220.00 per ounce. That is roughly $1,098,000 in value. The question is, what did you spend to get here? Hopefully you have kept track of all expenses for the IRS. And you know your four piles will yield over $4,000,000.

If you need more capacity, simply add another set of everything, bowls, impact, table, and all, to your circuit, and you are on your way.

What Can Go Wrong?

Of course, you have been assaying your tails off the table regularly, and they are just too high. You remember *"Heads Minus Tails = Recovery"*, correct?

The most likely problem is that the impact mill is over grinding your ore. The gold particles are being flattened, and the finer particles are simply floating past your bowls and table. The more clay you have in your ore, the higher the losses will be. The clay will help suspend the smaller particles.

The impact mill, is, in essence, creating flour gold by over-grinding your ore.

What to do? Take sample of your head ore to a soil testing lab, have it analyzed for clay content. Try for a coarser grind, say -80 instead of -100. Process some more ore, then assay again. If all else fails, you may have to go to a ball mill instead of an impact mill. Keep in mind that a ball mill will always retain some gold. When you clean it out, that black slimy stuff is accumulated gold and silver that can't, or hasn't made it through the discharge screen. Read more about this under the "Ball and Rod Mill" section in Chapter Ten.

The ball mill may be more expensive to operate and maintain, but it will do the job. Think about it. Impact mills slam the ore particles against the anvil(s) in the impact mill at a very high velocity. Gold, being malleable, will tend to flatten out. The clay is pretty much the kiss of death because it will help suspend the gold particles, preventing recovery by specific gravity.

The ball mill breaks the rock by smashing the rocks together, and against the balls. Some may flatten, and some may be trapped in the ball mill, but your recoveries will be higher. There are ball mills out there from small four and six footers to the monsters used at the major mines. Note that your gravel mill has produced an acceptable feed for a ball mill.

Chemically, you can depress fine gold with sodium hydroxide, NaOH, or caustic soda. You will probably need to permit for this, as well as any other chemical. Keep in mind that caustic burns are just as dangerous as acid burns. Read the MSDS. This is one method of getting around a high clay content in your ore. You won't be able to discharge this chemical, so plan on keeping it around until you can evaporate it, or otherwise legally dispose of the solution.

Note that if you add sodium hydroxide, you raise the pH of your water. Most flocculents used in cyanide operations will probably work for you, as they have a high pH as well. A quick test with a pH meter or at least pH paper would be good.

This is why you run pilot tests. Find and solve the problems before you invest in a lot of equipment that doesn't work. Off to the dump stocks.....

The Dump Stock and Mine Dump

The first issue to think about is where you will mill this ore. Most likely, you will be better off trucking the ore to a suitable location, where you have built a mill. Usually, these dumps are in remote areas, you might even have to build or widen a road to get to the dumps. Check out any right of way issues before you build a mill. You might not be able to get to the dumps with heavy equipment.

Again, this will be pretty much mine run ore, assume six to eight inch chunks of ore as the maximum. This time, build a conventional crushing circuit consisting of a primary jaw crusher, say 10" X 24", a secondary jaw crusher, say 6" X 18", and a large set of rolls. The primary crusher would crush to three inches, the secondary to one inch, and the rolls would take the ore down to ½ to ¾ inch or so, our ball mill feed.

Keeping in mind that the crushers and rolls all have different feed rates, and you don't want to spend all day clearing jams, you will need several feed hoppers and conveyors. The first one would feed the primary jaw crusher. The second one would take the discharge from the primary jaw and feed the secondary jaw. The third one would take the discharge from the secondary crusher and feed the rolls. The fourth one would take the discharge from the rolls and feed the ball mill.

The gravel plant isn't such a bad idea, after all. Think about what you would spend to get to this point.

Next you would install the ball mill. You need a large, thick concrete pad. A ball mill is very heavy by design, as well as incredibly noisy. You have managed to find an 8' X 10' ball mill, and have managed to meet the power requirements at the site. Keep in mind, the circuit is usually covered, so you have erected a steel roof to protect all the equipment from the elements. The ball mill discharge screen is set a 100 mesh, so everything coming out of the mill is -100.

Instead of using bowls, you decide on a double deck Wilfley (or Deister) roughing table. The ball mill discharge will go directly to the table, where a rough concentrate is generated. Off to the side, you have a Gemini lab table for cleaning up the concentrate, and producing smelter grade gold concentrate.

Or you could have the high grade (top) cuts off the Wilfley table going to a bigger Gemini table that runs when the other tables run, producing smelter grade gold as the plant runs. Be sure the tables are allowed to run clean before you shutdown.

Never, ever leave the gold from the Gemini table on site. Remove it daily. If you encounter some strange heavy metal coming off any cut on the Gemini table, take an ounce or so, and get a quick multi element spectrograph on the sample. Always know what you're dealing with. And again, *never, ever dispose of anything without an assay!*.

This time, we have eliminated the possibility of over grinding, and as we monitor and assay out tails, we know our recovery is where it should be. Keep track of weights, volumes and all that. Calculate your recovery, make sure it is where it should be based on your original samples.

What Can Go Wrong?

The discharge screen could get torn, or blow out. That's pretty obvious when it happens, you will have a lot of coarse rock coming out the ball mill discharge. Replace the screen, and try to find out what tore or destroyed the original one.

Your tables could be set up incorrectly. Follow the manufacturer's recommendations. Never set a table up on bare ground. The vibration will cause the table support legs to dig into the dirt, and you won't be anywhere near level anymore. Without being anchored, the table will try to "walk" and you'll lose level that way as well.

Try panning the tail off the tables. Pan tails from both. See gold? On one, or both? Make sure you have the tables set right, re-install if necessary. No gold? Check the ball mill discharge. Make sure you are providing enough water for the ball mill to flush out the ground ore. No gold? Check the feed. A certain amount of gold will be retained in the ball mill, but that should stabilize after an hour or so, and you should have gold showing in the ball mill discharge. If no gold shows in the feed, find out why. Re-assay the ore pile you are running.

Check your grind several times a day. Use a sieve, not your magic eyeball to check the grind. You shouldn't have the over grinding problem previously described. Your feed may be too slow, as well. Make sure the scoop that feeds the ball mill has plenty of material in the hopper.

The Placer Claim

There are many manufactures of portable placer machines out there. Your author's favorite over the years has always been Goldfield. Google them, they have a machine for every size operation, from hand fed right on up to the big plants. They are also the owner and inventors of the Goldfield riffle. There are also a lot of other manufacturers, check them out as well. Make an informed decision.

Maybe the manufacturer, or equipment dealer will loan or rent you a unit to try. Or maybe you can take your samples to the manufacturer for testing. It never hurts to ask, and then you know what you will get. Get references if you can, make the phone call, and see what their customers say. Research several different manufacturers.

Another tip is to use a feed hopper to feed whatever type of machine you decide on. A steady, constant feed really helps the recovery. Be smart and grizzly out the large rocks before the feed hopper. You might want to put a carpet mill on the tail end of the sluice, especially if your water is dirty.

Start your trench, or pits, and try to place the same amount of material in each barrel. Suppose three feet fills a barrel. Label the barrel, and dig the next three feet, and so on, until you (hopefully) hit bedrock. Try to dig the same size pit as close to bedrock as possible. That's where the good stuff is.

Continue on until you have sampled the property. Remember, mark each spot well, you might have to return to that exact spot a year later. Be sure to take lots of pictures. Will you find it?

Find yourself a suitable placer machine, such as Goldfield's "Prospector" or "Explorer" and a location with clean water. If you have to have stilling ponds, get some stock tanks at the local farm supply house.

Once you have your machine in place, bring your barrels to the site. Ideally, you would have done this at the site, but sometimes you can't.

Weigh the barrels. A simple platform scale will work, and at the end of the process, you'll have to know how much material you have processed. Also make yourself a cubic foot box, and from time to time, weigh a cubic foot. Once you know the weight of a cubic foot, total the weight for each hole. Divide the total weight of the holes by the weight for a cubic foot. That gives you the total in cubic feet, which you would divide by 27 to get the total cubic yards.

Run the barrels one at a time. Wash the concentrate from the recovery device, probably a sluice box, into a five gallon bucket. Label the bucket clearly, and if you have a small nugget or two, put them in a sandwich bag, and put them in the bucket, as well. Make a fast pass over the oversize from the plant with a metal detector to check for large nuggets.

If you want to check for a little better fine gold recovery, extend and widen (taper) your sluice box, and put ribbed carpeting in the extension. Widening the extension a few inches at the lower end creates a low pressure area and will let the fine gold settle. The ribbed carpet is manufactured by 3M under the name "Nomad". Some people will also run ribbed carpet and miner's moss together, a matter of preference. A quick search of the web will show you examples of both.

If you extend the sluice, just wash the contents of the ribbed carpet right into the bucket with the sluice box concentrate. Set that bucket aside, and go to the next barrel. And next. Until you are done for the day, or have processed all your barrels.

Now, you can process your concentrates, and calculate your recoveries.

A good way to process the concentrates is to classify them through a series of sieves, and put them across a lab sized (four foot) concentrating table or finishing table using clean water. The classification will keep like sized particles from pushing gold off the table by larger particles of rock. The closer in size, the better the separation of material by specific gravity will be.

Process the concentrate from each barrel separately, keep the gold from each barrel separate, dry it and weigh it on a balance (scale) that will read to a tenth of a gram, or a tenth of a troy ounce, either will work. Carefully weigh the gold from each barrel, *and write this down.* After processing all the barrels from each pit, add them together. *Write that down.* Do the next pit the same way. When you're all finished, average all the pits together.

Once you have processed all the barrels, you will know what you have, and where it is, and most importantly, *what you can expect to recover.* Once you know how many yards you processed, and how much gold you recovered, well, there you have it.

Technically, recoveries are measured in grams per yard. There are 31.1034 grams in a Troy ounce, and 12 Troy ounces in a Troy pound. Many placer miners consider anything above 10 grams per yard high grade. The really big operations sometimes survive on considerably less, but move hundreds or thousands yards per day.

Suppose you have done your pilot work, and are now running 100 yards an hour, and recovering 10 grams of gold average. That means you are recovering 1,000 grams per hour, or 32.15 Troy ounces or so per hour. Assume your gold is 80 fine, so you would be recovering 25.72 Troy ounces of fine gold. Of course, if you are producing nuggets, well, the sky is the limit. Nuggets, depending on the size, can sell for many thousands of dollars per ounce, as opposed to spot price for fines.

Note that Mother Nature never evenly distributes the gold. That's why you do test trenches and or pits. Then you will know which areas producing the most, or the least gold and at what depth.

If you are wondering about overburden, run your pits shallow, say three feet or so, process, then go down another three feet, process, and so on. Watch for a layer or layers of caliche, heavy clay (hard pan) and such. This would be false bedrock. Always check under the layer of clay or caliche, sometimes the gold is under the layer. The best values are normally on bedrock, real or false.

If you have trouble with clay in a trommel, most machines are made to accommodate some steel rods in the barrel, or high pressure water jets to break up the clay. There are also some chemicals that help, but chemicals can open another can of worms if you use them. Permitting is hard to do, and expensive.

When you calculate your recovery, watch out for the nugget effect. It will skew your results. That's the occasional nugget. If you are consistently recovering similarly nuggets, well, that's part of the package, and accordingly, can be calculated into the recovery. If you recover a real bruiser nugget, and it's an oddity, call it a high flier, and don't base your budget on it.

All placer machines come equipped with a nugget traps, but that is for the nuggets that have been classified in size by the punch plate, and will go through whatever recovery device you have such as a sluice box.

Always check the oversize out of your machine with a metal detector. Trommels use punch plate with a certain size hole to classify the feed. If the nugget is larger than the hole in the punch plate, the nugget goes out with the oversize. If this is happening, get another trommel with larger holes in the punch plate. If the nuggets are too large, you'll have to put a system in place to recover them. Or perhaps your machine will have a nugget trap *ahead* of the trommel portion of the plant.

To get your permits, you may be required to save the top soil for later reclamation. Usually, an earthmover is used to scrape off the top foot or so of soil, and pile it where it can be recovered for the reclamation process.

Some placers have up to a hundred feet or more of overburden over the values. This is why Caterpillar invented the D-10. The overburden is pushed aside, the valuable placer material underneath is processed, and the overburden is pushed back. The bedrock is usually processed manually to insure a complete recovery.

After cleaning your concentrate, it is considered in mining circles that anything that will go through a window screen (approximately -12 mesh) is "dust" or "fines". From that point, you should read "How to Smelt Your Gold and Silver". Gold buyers were paying dredgers 50% on their fines because the fines had some black sand mixed in.

The purpose of that book was to teach people how to clean up the fines with a simple smelt, and get a decent price for the gold. The smelt also eliminated the black sands. The book also deals with many other materials. Let's face it, buyers pay more for gold bars than gold fines with visible black sands.

In closing the placer portion of this chapter, keep in mind there is no 24 karat or 99.9 fine gold that comes out of the ground. Most is 80 or 90 fine, and must be refined to a purity of .999. So if you have a 10 Troy ounce bar, you get paid for 8 ounces of gold, and hopefully 2 ounces of silver.

It is always better to deal with a refiner if possible. They will give you a percentage of Hallmarked gold in return for your gold, if you desire, or pay you for the gold. Check their policy on shipment, assaying and refining charges. Remember as well that a Federal license is required to Hallmark gold or any other precious metal. Your author dealt with David H. Fell and Sons in The City of Commerce, CA and was extremely satisfied with them.

What Can Go Wrong?

Clay, for one, and theft, for another. The clay issue has already been discussed, and the theft issue can be solved with a little common sense. Maybe a security camera? Or how about a locking cover on the nugget trap and sluice box?

Suppose you have a good day, and left, returning the next morning to find a flash flood washed your plant away? Or some nefarious type stole your plant? Stranger things have happened. What about good, old fashioned armed robbery? Can you see who's coming down the road? Do you trust your fellow man not to do this? If you do, you're going to get a really nasty surprise. Convenience store clerks are murdered for a bag of potato chips.

Think about these issues, and plan accordingly.

Old Mill Tailings

They set out there in the desert, in the hot sun and elements for a long time, maybe even a hundred years oxidizing more and more. Further oxidation comes from the residual chemicals in the tails from the original leach process. Normally, this would be sodium hydroxide, if cyanide was used. The tails will be a strong caustic (alkaline) and the dust from the tails can be hazardous to breathe. Protect yourself. Think "respirator".

Old mill tails are great to work with, since the grinding is done. Saves you a ton of money. Take a few of your samples and do a screen fraction analysis, or have one done. This will give you two vital pieces of information. First you will find out what the original grind was, and second, the values in each mesh size as illustrated in the previous screen fraction analysis. Do this, if you do nothing else, this is need to know information.

Acme Assay Company
Assay Report - Gold Only - Screen Fractions #4

John Smith Mine   123 Easy Way Rd.   Gold Mine, AZ

Sample ID:	Sample Name	Au, Oz/Ton	Ag, Oz/Ton
1	-10 to +12	0.001	N/A
2	-12 to +15	TR	N/A
3	-15 to +20	0.002	N/A
4	-20 to +30	0.001	N/A
5	-30 to +40	0.089	N/A
6	-40 to +50	0.150	N/A
7	-50 to +60	0.250	N/A
8	-60to +70	1.400	N/A
9	-70 to +80	1.350	N/A
10	-80 to +90	0.580	N/A
11	-90 to +100	0.200	N/A
12	-100 to +110	0.008	N/A
13	-110 to -120	0.002	N/A
14	-120 to +130	0.001	N/A
15	-130 to +140	0.001	N/A
16	-140 to +150	0.002	N/A
17	-150 to +160	0.001	N/A
18	-160 to +170	TR	N/A
19	-170 to +180	0.001	N/A
20	-180 to +190	TR	N/A
21	-190 to +200	0.001	N/A
22	-200 to -250	0.001	N/A
23	-250 to +300	NG	N/A
24	-300 to +350	NG	N/A
25	-350 to -400	NG	N/A
26	-400	NG	N/A
Note:	NG = No Gold TR = Trace		

Assayer: Leroy Jones   2/14/2014

Hopefully, you have panned 10 lbs or so, and have recovered some sort of concentrate in your pan. What did it assay? If you weighed the sample before you panned it, and then dried and weighed your pan concentrate, you should have a rough idea what the concentration ratio will be. Don't be surprised if it's something like 100 to one. In other words, 100 tons of ore will produce one ton of concentrate.

Basically there are two possibilities with material of this nature. If it doesn't pan well, that will make gravimetric concentration unlikely. The next option is to re-leach, probably on a leach pad you will have to construct. Prior to re-leaching, you will have to *agglomerate* the ore. Agglomeration is necessary because the fine particles of the old tailings won't let the leach chemistry percolate through the heap.

Agglomeration is accomplished by mixing the wet tails with Portland cement, lime, or other ingredient in a rolling metal tube. The end result is basically a glop of fairly stiff mud four or five inches long that look like scones. The glops of mud are allowed to dry, then placed on the leach pad. This allows percolation in the re-leach. Re-read the previous information on heap leaching.

After reviewing Screen Fractions #3 above, You will notice the original mill was doing a -100 mesh grind, typical for a cyanide operation. Since most of these tailings will concentrate using gravimetric methods, we will start there. You do have a few problems to deal with, however.

The tailings have been in place for a hundred years or so, they are going to be pretty hard, but not rock hard. You need to break them up, or actually, break them back down to the original grind. How about a wide set of rolls, and a few feed hoppers and conveyors? Find yourself a set of rolls 12" or 14" wide. The rolls will break up the clods quite well. If a tramp rock goes through them, no harm will be done.

Next, a vibrating screen deck to remove the oversize stray rocks and gravel. From there, your re-ground and screened material heads for a hopper, ready for gravimetric processing.

But wait...You might be working too hard. How about a nice Goldfield Alaskan trommel plant? This type of plant is made to do exactly what has been previously described. You might want to change the screen on the trommel, down to 10 mesh, since you're after fine material, and the plant can be had with Pan American jigs, and various other fine gold recovery options.

The main drawbacks would be the availability of sufficient water, and the fact the smallest production plant will process 25 tons per hour. Goldfield also makes cleanup tables and other handy gizmos. Also, keep in mind the plants are portable, in the sense that they can be broken down and moved. They do, however meet your requirements for a pilot run. And they are common, parts are available, there are even smaller plants that would work.

Investigate this if you have old mill tails.

What Can Go Wrong?

First and foremost, not having enough water. You have to have enough water, or you will simply be making mud pies. The other thing that can be an issue is organic growth. Got trees? Obviously, they have to go, as well as tumbleweeds. All the brush, and all the crap that has accumulated over the years will have to be removed.

The water you use won't be able to be discharged, either. The residual sodium hydroxide in the old tails will raise the pH of the water to the point it will be illegal to discharge. If you wind up re-circulating the water, it will have to be screened to prevent access by animals, livestock, children and so forth.

Go to Wal-Mart, or a pool supply house and buy some pH test paper. Fill a quart jar half full of the water you will use, or just tap water. Check the pH of the water. Add a cup of the old tailings, and stir well. Wait an hour or two, and check the pH. Your water should have been around a pH of seven, which is neutral. If the old tails raise the pH above eight or so, and you re-circulate, the pH of the water will get higher and higher the longer you use the water.

Any residual cyanide should be long gone, since sunlight and heat decompose cyanide, and a sodium hypochlorite kill should have been done on the tails, but who really knows? Do the procedure above, but after checking the pH, put the water through a coffee filter, saving the liquid. Do this with the same water three or four times. Double the filters on the last rinse, get the water as clear as you can. If you happen to have a little dilute silver nitrate solution around, say 5% or 10%, put one drop in the water. The silver nitrate will turn white in the water. If there are low concentrations of cyanide present, it will digest the silver, and the water will turn clear.

You can also take the solution (water) to your local lab, they should be able to do a simple cyanide titration. If you even remotely think cyanide is present, get professional testing done, don't take a chance. Cyanide is a Class Six poison, and will kill you really quick.

Find some rusted out old barrels of chemicals buried in the tails? *Don't touch or disturb them in any way.* Get with a Hazmat crew, and get them on site to determine exactly what they are. Your author has encountered this. Been there, done that, and got the gray hair to prove it. It is incredible and scary what has been buried in old tailings.

A few years back, at an operating mine in Nevada, the operators were caught burying 55 gallon drums of mercury. Loader buckets of mercury were also being dumped in the tails. So, you see, anything is possible. Be prepared to deal with this.

Chapter Twenty Two

Building A Mill

Location

Location, location, location, they say in the real estate business. Since milling is an industrial process, location is the key. You can understand why most mills are located at some remote, out of the way place, typically right on a mine site. Think about appearances. The motoring public doesn't want to see what they perceive as eyesores. The fact that their stone washed jeans, the wiring in their houses, and even the jewelry on their fingers are the products of mining doesn't enter into the equation. And then, irregardless of economic benefit, there are the Nimbys (Not In My Back Yard). These are the same morons that buy a house at the end of an airport runway, then complain about the airplanes overhead. Or build a house on a known, active fault, then whine when an earthquake wipes their house off the planet. The examples of the lack of common sense in the human race can easily fill this book, however would serve no useful purpose.

If you have claimed a mill site on Federal land (BLM, Forest Service, etc.) You'll find out quickly there are a separate set of rules in this case. To get a feel for how the Federal agencies perceive mines, mining, and miners, you should read General Technical Report INT-GTR-35, "Anatomy of A Mine From Prospect to Production". This 69 page booklet is downloadable on the internet as a PDF document, and is the first publication anyone contemplating an operation on Federal land should read.

How many acres do you need? Always create a buffer zone, wherever you build. Give yourself all the room you think you need for the duration, then double it. The further from anyone, the better. Choose your site carefully. Remember that you can claim a mill site in conjunction with your claims, and in days of old, a water right claim came with a mill site. Mill sites were of a specific acreage, so check this out, as Federal Law changes daily, it would seem.

Old Mining Claims

Oh yeah, and there is this little trap to watch for. You think that since the material you are processing came from this old expired claim, you can file your own claim, and put in a mill. After all, if it came from the claim, you can leave the tailings on the claim. Sure you can. Be prepared for a hefty reclamation bond, and if the old claims have old shafts on them, you are a sitting duck. Every government agency will cheerfully take the appropriate fees, and *assign a portion, or all, of the cost of reclamation for the prior work to you, as the current claim holder.* Don't worry, they will notify you sooner or later. Think about that. What does it cost to fill and cap a hundred foot shaft? Or seal and old tunnel and cap it to prevent access forever?

There are idiots that will walk into an active mining operation, and commence to explore all the tunnels and shafts on what is now your claim. No, they do not seek permission first. They believe that if you are mining, there is something valuable in there, and they will go try to take something, which may even be your equipment from the site.

There are instances every year where some idiot will walk up to the very edge of a collar to a shaft to look in, or drop a rock in to see how deep the shaft is. A few are standing on a collar that is undercut, and the ground under their feet sloughs off, and they quickly find out how deep the shaft is when they hit the bottom. The problem is, if the wall rock from the old shafts are decomposing, the walls of the shaft shed rock constantly.

Going after the body, living or dead, is incredibly risky. Most Agencies no longer risk the recovery. With the advent of remote video, rescuers see a body, or watch the pitiful death of the "victim" in real time.

Don't worry about the video, you'll get to watch it in a courtroom when you are sued for wrongful death, since as the "owner" you are liable. The lawyers will go after the deepest pockets first, and well, you're next! And you must remember that the idiot achieved Sainthood, and had an earning capacity of millions per year before he dropped into your open shaft.

Overseas, there are some countries that will sue you if your employee was killed in a car wreck on the way home from work. Or whatever. You are liable for the lost earnings. This is a cold hard fact. Be prepared to deal with it. Some of the stories are incredibly hard to believe, but they are mostly true. Never doubt the greed of a weasel attorney.

Private Leases

Another common mistake is to lease private ground for a mill site. All is well and good until you find out the property you leased is zoned agricultural. You need industrial zoning. *Check the zoning!* People get real testy about industrial operations where previously, there were cattle grazing, or alfalfa growing. Now there's a dusty, noisy rock crusher running. What do you intend to do with the tails? Will solutions be dumped on the ground, possible contaminating the water? God forbid a tailings pond, or any chemistry. And the extra traffic is just unbearable to the neighbors. What was that word? Oh, yeah, Nimby.

Don't set down with a property owner and scribble a few lines on a paper. Retain a competent mining attorney, and have the lease at least reviewed prior to signing. This lease will be mostly about liability on the property. A simple gravimetric circuit may be possible, but again, you need at least I-1 (Industrial 1) zoning.

One or more persons may "partner" and provide the property for a few "points" of your return. Or maybe you have a "Cripple Creek lease, where the property owner gets 10% of the "net smelter return".That's all well and good until something goes wrong. What will you do if one of your employees fall into the crusher? Or removes their fingers in a jammed set of rolls?

Insurance Requirements

Make sure you do the appropriate payroll deductions, such as State Workers Compensation, (Insurance) Unemployment Compensation, and so forth. If you have enough employees to come under the purview of MSHA (Mine Safety and Health Administration) as opposed to a State OSHA (Occupational Safety and Health Administration), these folks will drop by and hand you a stack of citations for things you never thought possible. The Workers Compensation will take care of the issues when Leroy falls in the crusher or removes his fingers in the jammed rolls, but OSHA or, if applicable, MSHA will investigate the "accident" and deal with you, the operator, in the manner they see fit. It won't be pretty, so do your homework and think through the safety issues. And yes, industrial accidents kill and injure a lot of people every year. Know the rules, and play safe.

Seasonal Considerations

Are you going to mill year around? What are the temperature extremes? Are you leaching with chemistry? Typically, chemistry becomes less efficient the colder the weather. If you are using chemistry, and in the southwest, you can take advantage of the climate by simply painting your tanks flat black. Or flat dark green. This way, you get free heat. Remember that some chemistry decomposes at extreme temperatures.

Think about annual rainfall, snow and such. Do you need to be indoors? In the southwest, there is a monsoon season, and north, well, there's winter. In Alaska, back in the day, placer miners would erect a metal building over a stream. They would live on one side of the building, and process the other side. When they hit bedrock, they would move to the completed side, and process the other side. Come spring, they would remove the ends of the building, and let spring thaw pass through. They would then move the building up or downstream for the next season.

Think about sheltering your operation. After all, if your operation is shut down for weather, well, no production, no dough. Simple as that. A large metal building can suffice, even with dirt floors. Concrete foundations will be necessary for large pieces of equipment, such as ball mills, and so forth. It might be cost effective to go ahead and pour the floor while you have the equipment to do so onsite.

The point is, be aware of the climate. Do you have flash floods? A placer operation can be swiftly devastated by a flash flood, and flash floods are quite common. An operation on a riverbank or in a river valley can be at risk. What was the water level during the last 50 or 100 year storm? Is there an old earthen dam upstream? Don't let your investment wash away.

In the high country of Colorado, above the timberline, mines shut down for the winter. In the spring, many portals have a thirty foot or so ice plug that has to be mined out before the operation can resume. Wherever you may be, and whatever your operation, the climate and seasonal considerations are important. You may have to plan for extended shutdowns of a month or more.

Water

Water again. Seems like you always have to pay attention to water. How much, what quality, the treatment before and after use, and on and on. The simple fact is, water is a deal breaker. What type of operation dictates all the parameters associated with water, and if you can't get the required amount, well, your operation will be over with before it starts.

All states require well permits, and with the permit comes all the inquiries one would expect. Especially what the water will be used for, and if it's returned to the "resource," how will it be treated? Be prepared for all of this, and plan on a state agency putting your operations plan under a microscope. If you are overseas, and you buy a lawyer cheap and secure a permit, keep in mind the indigenous population. These days, the locals are well aware of chemical damage and such processes as amalgamation.

Major mining operations in foreign countries have been shut down for contamination. On of the more recent debacles was in Peru, where the waste from a mine was being dumped in a lake. The waste was high in arsenic. Bad plan, and the company didn't realize that there are always local environmentalists watching, wherever you may be.

Be realistic when you consider your water needs, and plan for the future. Remember that *water quality* is important as *water quantity*.

Equipment

This is where the flow sheet and icons from Chapter 13 come into play. At this point you should know exactly what equipment you need, and have a flow sheet drawn up for your circuit. Your flow sheet should have been established from the hard science you received from assays, multi element analysis and so forth.

Why would you spend $30,000 on a magnetic separator (or whatever) that you don't need? Don't fall victim to some used car (equipment) salesman when you are securing the equipment for your mill. Make sure you have enough floor space for your equipment, and some method of climate control for your employees.

Yes, that's snow, and plenty of it.

Chapter Twenty Three

Help

If you have managed to follow the information in this book this far, you should have a real grasp on what you are doing. Usually, issues arise due to the failure to follow the steps. If you decided you couldn't afford the analytical work, or perhaps the need for pilot testing, well, maybe you should go do something else.

Your circuit isn't performing as you predicted. You have pursued all the analytical information, and went from pilot to production. Do a forensic analysis of the circuit. Assay each step of the process, starting with the head ore through each step of the mill circuit to the tails.

Suppose you are running an agitated leach operation with thiourea. Head assay is good, tail assay is the same as the head assay. Assuming you have paid attention to your chemistry, look for *any exposed metal* in the circuit. A short piece of iron or aluminum pipe will cause your values to precipitate, and go to tail. Cyanide is less sensitive, and will work in black iron with no problems. The point is, you have to understand your process, and your chemistry, and your ore.

Check your grind. If you ran the tests previously described, you should know the point of mineral liberation. If, for one reason or another, your grind has slipped or failed, this would explain a higher tail assay. Something as simple as a screen failing can cause oversized particles to enter your circuit, and cause high tail assays. Inspect your equipment.

Books

As you will recall, there were considerable books listed in Chapter One. Review this literature, and understand that ore bodies change at various depths. Did you hit a transition zone in your ore body? Are you seeing a transition from oxide to sulfide? Are you seeing iron pyrite in your ore? You should know at this stage how sulfides will react in your circuit.

Another element that can sneak up on you is Tellurium. Or Selenium. Go back and review your original spectrographic analysis report, and run another analysis. Compare the two reports, see what has changed.

If you are mining, it is not uncommon for the ore body to alter the deeper you go. Simple compounds such as carbonates can interfere with your process as well. If you are processing an old tails dump, the same thing happens. Usually the tails on the top surface of the dump were the last mined. Hopefully, your recovery system was based on a good representative analysis.

Mining Professionals

By now, hopefully, you have had contact with a mining engineer, or a metallurgist. Either one can probably be of assistance to you with whatever problem you are having. If you are running a chemical leach circuit, it is a really good idea to know all the parameters of the chemistry you are using. For example, if you are using cyanide, and haven't established and maintained protective alkalinity, you aren't reading this because you are most likely dead.

If you are running a placer operation, or a hybrid circuit with free gold, a simple gold pan can provide a lot of answers. Just working your way through the circuit and panning after each step will tell you a lot. A little common sense goes a long ways. Obviously, this won't work with micron gold. In this case, you'll need to run some assays, or rely on a technical professional.

The Author

Yes, you can call. It is almost impossible to diagnose a circuit without being onsite. So, probably not too much help. Also, the questions you will have to answer will decide the relationship between you and the author. If you skipped all or some of the steps outlined in this book, talking with you will be a waste of time because you did not understand the basic premises outlined in this book. Also, make sure you have read the book, and have access to the internet before you call. Have your assay reports handy, as well as your spectrographic analysis. Be prepared to discuss your particle size distribution, and your circuit step by step. And safety, among other things.

After a point, *be prepared to pay for the time you use.* And yes, e-mail works just as well. The contact information is in the front of the book.

Gold and silver bars from your author's work.

Thank you for your purchase of this book!

Glossary

A good online resource for mining terminology is <u>www.infomine.com/dictionary/</u>**.**

Acid Rain- Precipitation containing acid-forming chemicals, chiefly industrial pollutants, that have been released into the atmosphere and combined with water vapor: ecologically harmful.

Acid- A substance having a pH value of less than 7. See pH.

Activated Charcoal- (or carbon) a form of carbon having very fine pores, used chiefly for adsorbing gases or solutes, as in various filter systems for purification, deodorization, and decolorization. Generally made from coconut shell by burning in a reducing atmosphere.

Adsorption- The process by which an ultra thin layer of one substance forms on the surface of another substance.

Alkaline- A substance having a pH greater than 7. See base, or basic. See pH.

Amalgam- An alloy of mercury with another metal or metals.

Amalgamation- The process by which mercury is alloyed with some other metal to produce an amalgam. It was used at one time for the extraction of gold and silver from pulverized ores, but has been superseded by the cyanide process.

Analog- Displaying a readout by a pointer or hands on a dial rather than by numerical digits.

Analysis- The ascertainment of the kind or amount of one or more of the constituents of materials.

Anhydrous- Dry, all water removed, especially the water of crystallization.

Aqua Fortis- Nitric Acid, HNO_3.

Aqua Regia- A mixture of nitric and hydrochloric acids used to dissolve precious metals.

Arsenic- A grayish white element having a metallic luster, vaporizing when heated, and forming poisonous compounds Symbol: As.

Ash- the powdery residue of matter that remains after burning. The process of burning a material to create ash.

Aspirate- to draw in by suction, as used with Atomic Absorption Spectrophotometer.

Assay Ton- A specific weight related to the number of grams in a short ton. An assay ton is 29.1667 grams.

Assay- To analyze (an ore, alloy, etc.) to determine the content of gold, silver, or other metal.

Attrit- The process of wearing loose particles from a material, such as activated charcoal, by gently agitating particles against each other in water.

Bag House- A structure containing filter media to remove contaminants from air, usually dust.

Baking Soda- Sodium Bicarbonate, $NaHCO_3$.

Balance- An instrument for determining weight.

Base metal- Any metal other than a precious or noble metal, such as copper, lead, zinc, or tin.

Basic- A substance having a pH greater than 7. See alkaline. See pH.

Bead- A small ball, or bead, of precious or noble metals remaining in a cupel after cupellation. Part of the fire assay process.

Black sands- Magnetite or Hematite. Forms of iron or iron compounds recovered as impurities in the placer mining process.

Bond Work Index- This procedure is used to determine power consumption in crushing and grinding to the feed and product size distribution.

Bone Ash- A white ash obtained by roasting, or calcining bones.

Borax Glass- An important flux ingredient made by calcining hydrated sodium borate, $Na_2B_4O_7 \cdot 10H_2O$.

Borax- A white, water-soluble powder or crystals, hydrated sodium borate, that is calcined to remove water and create borax glass. See borax glass.

Borosilicate Glass- See slag.

Bullion- Relatively pure noble metals, considered in mass rather than in value. Usually in bars or ingots.

Bureaucrat- An official who works by fixed routine without exercising intelligent judgment.

Calcine- to convert into calx by heating or burning. See calx.

Calibration- To set or check the graduation of a quantitative measuring instrument.

Calx- The oxide or ashy substance that remains after metals, minerals, etc., have been thoroughly burned.

Cap- To cover, or top off with a layer of flux, borax glass, or other dry reagent.

Carbon- A nonmetallic element found combined with other elements in all organic matter and in a pure state as diamond and graphite Symbol: C. See also activated charcoal. A reducing agent. Any source of carbon, such as flour or sugar can be used as a reducing agent.

Carbonate- Used to refer to specific types of ores, such as calcium carbonate, or calcite. Carbonates usually have a "CO_3" suffix.

Carbon Dioxide- A colorless, odorless, incombustible gas, CO_2, that is a by-product of smelting carbonate ores.

Carat- A unit of weight in gemstones. Not used for noble metals.

Carcinogen- Any substance or agent that tends to produce a cancer.

Cast- To form (an object) by pouring metal into a mold and letting it harden.

Celsius- Pertaining to or noting a temperature scale in which $0°$ represents the ice point (freezing) and $100°$ the steam point (boiling).

Cementation- A chemical process where an inexpensive metal, usually iron or aluminum, is used to cause a chemical reaction that will produce or precipitate a noble metal out of a solution that contains noble metal.

Chemical- a substance produced by, used in, or concerned with chemistry or chemicals.

Chloridize- To convert into chloride; applied to the roasting of silver ores with salt, preparatory to amalgamation.

Collar- The term applied to the timbering or concrete around the mouth or top of a shaft, or the beginning point of a shaft or drill hole, the surface.

Collector- A person or thing that collects. In a smelt, a large enough ratio of metal, that when molten, will collect other metals that are present.

Compound- A substance that is composed of two or more parts, elements, or ingredients.

Concentrate- to separate (metal or ore) from rock, sand, etc., so as to improve the quality of the valuable portion.

Condiment- Something used to flavor food, such as mustard, ketchup, salt, or spices.

Cons- Slang term for concentrate, see concentrate. The end result of the concentrating process.

Contaminate- To inadvertently make impure or unsuitable by contact or mixture with something unclean or bad. To pollute or taint.

Copper- A malleable ductile metallic element having a characteristic reddish brown color. Used in large quantities as an electrical conductor and in the manufacture of alloys, as brass and bronze Symbol: Cu.

Corrosive- Having the quality of corroding or eating away; erosive, such as acid vapors, or solutions.

Crucible Wash- A dry reagent that is lighter than other flux constituents when molten, and therefore is the last portion of the molten material to leave the crucible when poured.

Crucible- A container of refractory material employed for heating substances to high temperatures.

Crystalline- Of or like crystal; clear; transparent.

Cubic Feet Per Minute- A term used to describe the volume of air that is being moved in one minute. CFM.

Cupel- A small, cup-like, porous container, with a hemispherical depression to focus heat in the center. Usually made of bone ash or magnesite, and used in assaying, for collecting gold and silver from lead. The bone ash or magnesite absorb about 90% of the lead, the remainder is vaporized as lead oxide.

Cupellation- To heat or refine in a cupel. The process of removing lead from noble metals in a fire assay.

Cyanide- A salt of hydrocyanic acid, as potassium cyanide, KCN, or sodium cyanide, NaCN. To treat with a cyanide, as an ore, in order to extract gold.

Desulfurizing Agent- A dry reagent used in the fire assay or smelt to remove sulfur. Soda ash is a common reagent used for this purpose.

Dewater- To remove moisture from a slurry by various means, such as a thickener, belt or drum filter, or to remove water from a mine; an expression used in the industry in place of the more technically correct word, unwater.

Digital- Displaying a readout in numerical digits rather than by a pointer or hands on a dial.

Discharge Permit- A permit from a governmental agency allowing the discharge of a specified amount of a toxic or polluting compound from an industrial facility into the environment.

Dore'- An alloy containing gold.

Electrolytic cell- A container with an anode and cathode that a precious metal bearing solution is passed through. Low voltage is passed through the cell to remove the precious metal from the solution.

Elements- One of a class of substances that cannot be separated into simpler substances by chemical means.

Endothermic- Noting or pertaining to a chemical change that is accompanied by an absorption of heat.

Exothermic- Noting or pertaining to a chemical change that is accompanied by a liberation of heat.

Fahrenheit- Noting, pertaining to, or measured according to a temperature scale in which 32° represents the (freezing) ice point and 212° the (boiling) steam point Symbol: F.

Fineness- The proportion of pure precious metal in an alloy, often expressed in parts per thousand.

Fire Assay- An analytical process utilizing heat and dry reagents to quantitatively determine the amount precious metals in an ore. Considered to have a detection limit of .001 ounces per ton.

Firebrick- A brick made of fire clay.

Fire Polish- To repeatedly smelt with an oxidizing flux to increase the fineness of the precious metal content, usually gold.

Flammable- Easily set on fire; combustible.

Flocculent-Particulate in a solution coalescing and adhering in flocks. A cloudlike mass of precipitate in a solution.

Flour- The finely ground meal of grain, especially wheat. Used as a source of carbon, and as a reducing agent in the fire assay.

Flue dust- Dust accumulating in a flue, or ventilation system. May contain very high precious metal values.

Fluorite- A mineral, calcium fluoride, CaF_2, occurring in crystals and in masses: the chief source of fluorine. Also called Fluorspar.

Fluorspar- Calcium fluoride. See Fluorite.

Flux- A substance used to refine metals by combining with impurities to form a molten mixture that can be readily removed. Usually made from dry reagents.

Forceps- An instrument, as in pincers or tongs, for seizing and holding objects firmly, as in surgical operations. Giant tweezers.

Frother- A substance used in a flotation process to make air bubbles sufficiently permanent, principally by reducing surface tension.

Fumes- Any smoke like or vaporous exhalation from matter or substances, especially of an odorous or harmful nature.

Furnace- An apparatus in which heat may be generated, as for heating houses, smelting ores, or producing steam.

Fusion- The act or process of fusing or the state of being fused. To combine or blend by melting together; melt.

Galena- A common heavy mineral, lead sulfide, PbS, occurring in lead-gray crystals, usually cubes, and cleavable masses. The principal ore of lead.

Gangue- the worthless rock or vein matter in which valuable metals or minerals occur.

German Silver- Any of various alloys of copper, zinc, and nickel, usually white and used for utensils and drawing instruments; nickel silver.

Gold- A precious yellow metallic element, highly malleable and ductile, and not subject to oxidation or corrosion Symbol: Au. One of the noble metals.

Grade- A term used in the mining industry to denote values contained in an ore, or other product. There can be high grade, or low grade. To "high grade" can mean to steal.

Graphite- A soft native carbon occurring in black to dark gray foliated masses: used for pencil leads, as a lubricant, as a moderator in nuclear reactors, and for making crucibles and other refractories; also known as plumbago.

Gravimetric- A method of mechanical separation, usually with water, by specific gravity, or referring to specific gravity. Normally, concentrating tables, jigs, centrifuges, sluice boxes, or other devices are used for gravimetric separations.

Gyratory- A widely used form of rock breaker or crusher, in which an inner cone rotates eccentrically in a larger outer hollow cone.

Hallmark- An official mark or stamp indicating a standard of purity, used in marking gold and silver articles.

Head Assay- The assay of head ore, or the assay of the original material before any processing or treatment that may change the characteristics of the original material.

Heat Sink- An environment or medium that absorbs excess heat.

Homogenous- Well mixed or blended. Representative of the whole.

Hydrate- Any of a class of compounds containing chemically combined water.

Hydrofuge- A device that uses centrifugal force, in the presence of water to separate classified materials by specific gravity.

Hygroscopic- Absorbing or attracting moisture from the air.

IRS- The Internal Revenue Service. And they have no sense of humor.

Inert- Having little or no ability to react.

Ingot- A mass of metal cast in a convenient form for shaping, remelting, or refining.

Iodized- To treat, impregnate, or affect with iodine or an iodide.

Inorganic- Noting or pertaining to chemical compounds that are not hydrocarbons or their derivatives.

Inquart- Times four. To add a measured amount. Such as adding four times as much silver to a measured amount of gold to allow complete parting, or chemical separation.

Insulator- A material that absorbs or deflects heat.

Iron- A ductile, malleable, silver-white metallic element, used in it's impure carbon-containing forms for making tools, implements, or machinery. Symbol: Fe. A primary ingredient in steel.

Karat- A unit for measuring the fineness of gold, pure gold being 24 karats fine.

Kiln- A furnace or oven for burning, baking, or drying something, especially one for firing pottery, calcining limestone, or baking bricks.

Kilo- A metric unit of mass, one thousand grams. A Kilogram. Approximately 2.2pounds.

Lead Acetate- a white, crystalline, water-soluble, poisonous solid, $Pb(C_2H_3O_2)$ 2x $3H_2O$. Toxic by ingestion.

Lead- A heavy, comparatively soft, malleable, bluish-gray metal, sometimes found in its natural state but usually combined as a sulfide, as in galena. Symbol: Pb.

Lime- A white or grayish white, odorless, lumpy, very slightly water-soluble solid, CaO, used chiefly in mortars, plasters, and cements, in bleaching powder, and in the manufacture of steel, paper, glass, and various chemicals of calcium.

Limestone- A sedimentary rock consisting predominantly of calcium carbonate, varieties of which are formed from the skeletons of marine microorganisms and coral: used as a building stone and in the manufacture of lime.

Liquefy- To make or become liquid, with the use of high temperature.

Litharge- A yellowish or reddish poisonous solid, PbO, used chiefly in the manufacture of storage batteries, pottery, enamels, and inks. A very important ingredient in fire assay flux.

Manganese Dioxide- A hard, brittle, grayish white, metallic element, an oxide of which, MnO_2, is a valuable oxidizing agent: used chiefly as an alloying agent in strengthening steel. The natural ore of manganese is pyrolusite.

MSHA- (Mine Safety and Health Administration)

Malignant- As in tumor, characterized by uncontrolled growth; cancerous, invasive, or metastatic.

Material Data Safety Sheet- A document produced by a chemical manufacturing company to advise consumers of the hazardous properties of that chemical. Also known as an MSDS.

Mechanical error- An error resulting from mechanical handling, such as a dusting loss when finely divided materials are handled roughly, or poorly.

Mercury- A heavy, silver-white, toxic metallic element, liquid at room temperature, used in barometers, thermometers, pesticides, pharmaceuticals, mirror surfaces, and as a laboratory catalyst; Quicksilver. Symbol: Hg.

Mesh- An arrangement of interlocking metal links or wires with evenly spaced, uniform small openings between, as used in jewelry, sieves, etc.

Metallic- Of, pertaining to, or consisting of metal. Being in the free or uncombined state, such as metallic iron.

Mill- A mineral treatment plant in which crushing, wet grinding, and further treatment of ore is conducted.

Mine- An opening or excavation in the ground for the purpose of extracting minerals; a pit or excavation from which ores or other mineral substances are taken by digging; an opening in the ground made for the purpose of taking out minerals, or a work for the excavation of minerals by means of pits, shafts, levels, tunnels, etc., as opposed to a quarry, where the whole excavation is open.

Mine Safety and Health & Training- MSHA. The Federal Metal and Nonmetallic Mine Training, Safety & Health Standards as defined in 30 CFR, 46-48, 56-58, and 62.

Mining- The science, technique, and business of mineral discovery and exploitation. Strictly, the word connotes underground work directed to severance and treatment of ore or associated rock. Practically, it includes opencast work, quarrying, alluvial dredging, and combined operations, including surface and underground attack and ore treatment.

Mint- A place where coins, paper currency, special medals, etc., are produced under government authority.

Mold (Pouring)- A hollow form for giving a particular shape to something in a molten or plastic state.

Molten- Liquefied by heat; being in a state of fusion.

NAFTA- North American Free Trade Agreement.

Nepotism- Favoritism (as in appointment to a job) based on kinship.

Neutral- Exhibiting neither acid nor alkaline qualities, having a neutral pH.

Neutralize- To make (a solution) chemically neutral. To change pH.

NIMBY- An acronym meaning **N**ot **I**n **M**y **B**ack **Y**ard.

Niter- Potassium Nitrate or Sodium Nitrate. A strong oxidizer and important flux ingredient.

Nitric acid- A colorless or yellowish, fuming, suffocating, water-soluble toxic liquid, HNO_3, used chiefly in the manufacture of explosives and fertilizers.

Nomex- The trade name for a brand of fire proof garments.

OSHA- (Occupational Safety and Health Administration)

Opaque- Not allowing light to pass through.

Ore- A metal-bearing mineral or rock, or a native metal, that can be mined at a profit. A mineral or natural product serving as a source of some nonmetallic substance, as sulfur.

Organic- Noting or pertaining to a class of chemical compounds that formerly comprised only those existing in or derived from plants or animals, but that now includes all other compounds of carbon.

Osmium- A hard, heavy, metallic element, densest of the known elements, able to form octavalent compounds: used chiefly as a catalyst, in alloys, and in the manufacture of electric-light filaments. Symbol: Os. A Platinum Group element.

Ounces Per Ton- OPT. A term indicating the Troy ounces of noble metals in a short ton (2000 pounds) of ore.

Oxidation- The process of adding oxygen to a chemical reaction, usually by the addition of an oxidizer, or oxidizing agent.

Oxide- A compound in which oxygen is bonded to one or more electropositive atoms. A term used to denote a non-complex ore.

Oxidizing agent- A chemical compound that gives oxygen to a chemical reaction.

Part- To separate. Usually refers to the separation of gold and silver by wet chemical means.

Personal Protective Equipment- Personal safety equipment such as respirators, steel toe shoes, lab coats, etc. It is considered the individual's responsibility to use and maintain this equipment, which is normally provided at the workplace.

pH- The symbol for the logarithm of the reciprocal of hydrogen ion concentration in gram atoms per liter, used to describe the acidity or alkalinity of a chemical solution on a scale of 0 (more acidic) to 14 (more alkaline, or basic).

Pin Tube- A piece of glass tubing that has been evacuated of air, designed to sample molten metal. The tube usually has a blister at the end to be immersed in the molten metal. The blister melts at a lower temperature than the tube, allowing the vacuum to pull several inches of molten metal into the glass tube. After cooling, the glass is broken away from the metal, or "pin", which is considered a representative sample of the melt.

Placer- A natural concentration of heavy metal particles, as gold or platinum, in sand or gravel deposited by rivers or glaciers.

Placer Gold- Gold recovered from a placer mining operation.

Platinum Group Elements- Platinum, Palladium, Rhodium, Osmium, Ruthenium and Iridium.

Pollutant- Any substance, as a chemical or waste product, that renders the air, water, or other natural resource harmful or generally unusable.

Portal- The surface entrance to a drift, tunnel, adit, or entry. Or the log, concrete, timber, or masonry arch or retaining wall erected at the opening of a drift, tunnel, or adit.

Potassium Nitrate- A crystalline compound, KNO_3, produced by nitrification in soil, and used in gunpowder, fertilizers, and preservatives; saltpeter; niter. A strong oxidizer in fluxes.

Precious Metal(s)- A metal of the gold, silver, or platinum group.

Precipitate- To separate (a substance) in solid form from a solution, as by means of a reagent.

Preponderance- The fact or quality of being preponderant; superiority in weight, power, numbers, etc. More of.

Proprietary- Manufactured and sold only by the owner of the patent, trademark or process. Closely held.

Protective Alkalinity- A basic, or alkaline pH range used in cyanide operations, typically 9.0

to 10.5. Cyanide will not evolve hydrogen cyanide gas as long as this pH range is maintained, usually with sodium hydroxide.

Pyrolusite- A grayish black mineral, manganese dioxide, MnO_2, the principal ore of manganese.

Qualitative- Pertaining to or concerned with quality. Qualitative analysis normally will indicate what elements are present, but not how much of the element is present.

Quantitative- Being measured by quantity. Quantitative analysis will indicate precisely how much of a single element is present.

Quench- To cool suddenly by plunging into a liquid, as in tempering steel by immersion in water.

Readability- Pertaining to the accuracy and weight range of an instrument such as a balance, scale, or other weighing device.

Reagent- A substance that, because of the reactions it causes, is used in analysis and synthesis.

Reducing Agent- A substance that causes another substance to undergo reduction and that is oxidized in the process. A source of carbon, such as flour.

Refine- To bring to a pure state; free or separate from impurities or other extraneous substances.

Refinery- An establishment for refining something, as metal, sugar, or petroleum.

Refining- The process of bringing to a pure state. The process of separating impurities.

Refractory- A material that retains its shape and composition even when heated to extreme temperatures.

Refractory Ore- An ore that is difficult to fuse, reduce, or work.

Representative- A typical example or specimen. A small portion that represents the whole.

Residual- Pertaining to or constituting a residue or remainder; remaining; leftover.

Respirator- A filtering device worn over the nose and mouth to prevent inhalation of noxious substances.

Retort- A vessel, commonly a metal chamber with a long neck bent downward, used for distilling or decomposing substances by heat. A device for separating gold and mercury (an amalgam) from one another.

Roasting Dish- A refractory container, usually round, used for roasting (oxidizing) a mineral sample.

Salt- A crystalline compound, sodium chloride, NaCl, occurring chiefly as a mineral or a constituent of seawater, and used for seasoning food and as a preservative.

Saltpeter- Naturally occurring potassium nitrate, used in making fireworks, gunpowder, etc.; niter.

Salting- The fraudulent adulteration of a sample, for example, adding a small amount of gold to a sample to make it appear that the gold content of the rock is much higher than it actually is. Salting may be accidental, caused by the fortuitous segregation of rich mineral during sampling. Sampling methods are conducted to reduce chance segregation to a minimum.

Sand- The more or less fine debris of rocks, consisting of small, loose grains, often of quartz.

Self Contained Breathing Apparatus- SCBA- A breathing device that supplies compressed air, as opposed to a respirator, which filters air.

Shaft- A vertical or inclined opening of uniform and limited cross section made for finding or mining ore, or ventilating underground workings.

Shotted- Containing small, round metallic particles.

Silica- The dioxide form of silicon, SiO_2, occurring as quartz sand, flint, and agate: used chiefly in the manufacture of glass, water glass, ceramics, and abrasives. Also called silicon dioxide.

Silicosis- A disease of the lungs caused by the inhaling of siliceous particles, as by stone cutters or miners.

Silmanite- A refractory compound used to manufacture vessels for use at high temperatures.

Silver Chloride- A white powder, AgCl, that darkens on exposure to light: used chiefly in photographic emulsions and in antiseptic silver preparations.

Silver- A white, ductile metallic element, used for making mirrors, coins, ornaments, table utensils, photographic chemicals, and conductors. Symbol: Ag.

Slag Pot- A tapered, heavy metal container used to contain smelted metals. The taper allows for separation of the metal from the slag.

Slag- The more or less completely fused and vitrified matter separated during the reduction of a metal from its ore. Borosilicate glass containing the impurities from a smelt.

Slake- To cause disintegration by treatment with water.

Smelting- The process of fusing or melting in order to separate metal contained. To obtain or refine (metal) in this way.

Soda Ash- Sodium Carbonate, Na_2CO_3.

Sodium Bicarbonate- A white water-soluble powder, $NaHCO_3$, used chiefly as an antacid, a fire extinguisher, and a leavening agent in baking. Also called bicarbonate of soda, baking soda. Evolves large amounts of gas at high temperature, not considered a useful flux ingredient.

Sodium Carbonate- Also called soda ash. An anhydrous, grayish white, odorless, water-soluble powder, Na_2CO_3, used in the manufacture of glass, ceramics, soaps, paper, petroleum products, sodium salts, as a cleanser, for bleaching, and in water treatment. A valuable flux ingredient.

Sodium Chloride- See salt. A crystalline compound, sodium chloride, NaCl, occurring chiefly as a mineral or a constituent of seawater, and used for seasoning food and as a preservative.

Spall- To violently break or split off in chips or bits.

Spot- The daily fixed price of gold and other commodities.

Sprout- To spontaneously erupt. A phenomena of molten silver at high temperature and cooling.

Stack Permit- See discharge permit.

Static Pressure- The resistance to the flow of air through duct work or piping that must be overcome by a blower.

Sterling Silver- Silver having a fineness of 0.925, now used in the manufacture of table utensils, jewelry, etc.

Stoney- Resembling or suggesting stone.

Sulfide- A compound of sulfur with a more electropositive element or, less often, group. Such as Iron Pyrite.

Sulfur Dioxide- A colorless, nonflammable, water-soluble, suffocating gas, SO_2, formed when sulfur burns: used chiefly in the manufacture of chemicals such as sulfuric acid, in preserving fruits and vegetables, and in bleaching, disinfecting, and fumigating.

Sulfuric Acid- A clear, colorless to brownish, dense, oily, corrosive, water miscible liquid, H_2SO_4, used chiefly in the manufacture of fertilizers, chemicals, explosives, and dyestuffs and in petroleum refining. Also called Oil of Vitriol.

Surfactant- A surface active agent, a substance that affects the properties of the surface of a liquid or solid by concentrating in the surface layer.

Suspended- To keep from falling or sinking, as if by hanging. To suspend particles in a liquid.

Thallium- A soft, malleable, bluish white metallic element. Used in the manufacture of alloys and, in the form of its salts, in rodenticides. Extremely toxic in some forms. Symbol: Tl.

Thermal Shock- Stress to a refractory container, such as a crucible, caused by heating to extreme temperatures and cooling.

Tilting Furnace- A large furnace that tilts on an axis as it is elevated to the pour position by mechanical means.

Touchstone- A black stone once used to test gold and silver by rubbing them on it. Used to refer to a streak (color) test.

Toxic- Acting as or having the effect of a poison. Harmful to the human body.

Toxicity- The quality, relative degree, or specific degree of being toxic or poisonous.

Translucent- Permitting light to pass through but diffusing it so that objects on the opposite side are not clearly visible. Frosted window glass is translucent.

Trommel- A revolving cylindrical screen used in size classification of coarsely crushed ore, coal, gravel, and crushed stone. The material to be screened is delivered inside the trommel at one end. The fine material drops through the holes; the coarse material is delivered at the other end.

Troy Weight- A system of weights in use for precious metals and gems, in which a pound equals 12 ounces (0.373 kg) and an ounce equals 20 pennyweights or 480 grains (31.1035 grams).

Uniodized- Does not have iodine added.

Unslaked- Not treated with water, referring to lime.

Upgrade- To improve or enhance the quality or value of a precious metal bearing material by chemical or gravimetric means.

Vapor- A substance in gaseous form that is below its critical temperature. Usually toxic if produced a high temperatures.

Ventilation- Facilities or equipment for providing ventilation. The process of moving air through an enclosed area to remove vapors, fumes, or dust.

Viscous- Of a glutinous nature or consistency; sticky; thick; stringy; adhesive.

Washing Soda- See sodium carbonate.

Winze- A subsidiary shaft that starts underground. It is usually a connection between two levels.

Zinc- A ductile, bluish white metallic element: used in making galvanized iron. brass, and other alloys, and as an element in voltaic cells. Symbol: Zn.

Useful Conversions

One Troy Ounce = 31.1035 Grams. One Troy Pound = 12 Troy Ounces
Or 480 Grains
Or 20 Pennyweight
Or 1.0971 Avoirdupois Ounces

One Avoirdupois Ounce = 28.3495 Grams. One Avoirdupois Pound=16 Ounces
Or 437.500 Grains
Or 18.2292 Pennyweight
Or 0.9115 Troy Ounces

1% = 10,000 PPM (PPM = Parts Per Million)
1 PPM = 1000 PPB (PPB = Parts Per Billion)
1 PPM = .029166 Troy Ounces Per Short Ton
One Short Ton = 2000 Avoirdupois Pounds
One Short Ton = 29,1666 Troy Ounces
One Metric Ton = 1000 Kilograms = 2204.6 Pounds
One Troy Ounce / Short Ton = 34.2857 Grams / Metric Ton or 34.2857 PPM

Approximately 244 gallons equals one ton. (Note: This depends on the specific gravity of the solution.) For Lab purposes, 1 milliliter = 1 gram, or 100 milliliters = 100grams.

Head assay - Tail assay = Recovery

If you need to anticipate recovery of a leach test, record the weight of your sample you are leaching, then divide it into a ton, and multiply the result by the amount of metal recovered. For instance, you are doing a 20 Lb. Leach test. 2000 / 20 = 100. If you recovered 0.1 grams out of your pregnant solution, your anticipated recovery would be 100 X 0 .1, or 10 grams per ton. The conversions above show that one troy ounce is 31.1035 grams. Divide 10 by 31.1035, and you have the anticipated recovery in ounces per ton, .32 ounces per ton.

To calculate the precious metals in solution, convert everything to Troy ounces per ton. Then measure the volume of the solution you have. Suppose you have 50 gallons of solution that assays 0.1 Oz/Ton. Divide 244 by 50 = 4.88. Multiply 0.1 X 4.88 = .49. This means you have almost a half ounce of metal. You can do the same with milligrams.

-Index-

www.ingramcontent.com/pod-product-compliance
Lightning Source LLC
Chambersburg PA
CBHW082104210326
41599CB00033B/6584